Archaeology and Entomology in the Eastern Mediterranean

Archaeology and Entomology in the Eastern Mediterranean

Research into the history of insect synanthropy in Greece and Egypt

Eva Panagiotakopulu

BAR International Series 836
2000

Published in 2016 by
BAR Publishing, Oxford

BAR International Series 836

Archaeology and Entomology in the Eastern Mediterranean

ISBN 978 1 84171 129 4

Volume editor: John Hedges BSc(Hons), MA, MPhil, FSA, FSAScot, FRAI, MIFA

BAR Publishing is the trading name of British Archaeological Reports (Oxford) Ltd. British Archaeological Reports was first incorporated in 1974 to publish the BAR Series, International and British. In 1992 Hadrian Books Ltd became part of the BAR group. This volume was originally published by John and Erica Hedges Ltd. in conjunction with British Archaeological Reports (Oxford) Ltd / Hadrian Books Ltd, the Series principal publisher, in 2000. This present volume is published by BAR Publishing, 2016.

Printed in England

BAR
PUBLISHING

BAR titles are available from:

BAR Publishing
122 Banbury Rd, Oxford, OX2 7BP, UK
EMAIL info@barpublishing.com
PHONE +44 (0)1865 310431
FAX +44 (0)1865 316916
www.barpublishing.com

Abstract

Fossil insects are examined from the excavation of Late Bronze Age Akrotiri, Santorini, Greece, and from an Egyptian site of Roman age, Mons Claudianus. Insect samples from ancient Egyptian botanical material stored in the Museum of Liverpool and insect remains found in coprolites from Tut'ankhamun's Workmen's village are also discussed.

The first part of the study discusses the methodological problems encountered, as well as the preservation of the material studied. A brief summary of the archaeological record of pests is given in the next chapter. As most of the insect material studied was recovered from stored products, the accounts of Classical writers on foodstuff storage are discussed in detail. The section on storage is an effort to present the continuity of use of traditional practices to reduce the risk of crop loss in the storerooms.

The second part is a study of the archaeoentomology of the Bronze Age settlement excavated on Santorini. Background information on the main site is given. The insect fauna of the settlement is considered, most of it recovered from the West House at Akrotiri. The discovery of a cocoon of a wild silk moth, found on the site, is discussed, together with a summary of the relevant literary record, general information on sericulture and further iconographic evidence from the Late Bronze Age Aegean. The discussion of a model hive leads to a survey of early evidence for beekeeping. Entomological work on ancient Egyptian stored product faunas was also undertaken to provide comparative material to the Akrotiri finds.

A brief summary of trade in perishable material, as recovered in the archaeological record, and its importance for the reconstruction of past cultures shows the value of archaeoentomological research. The conclusions of the work undertaken and recommendations for further research are offered in the final chapter.

Acknowledgements

This research was undertaken with financial help from the British Federation of Women Graduates, International Federation of University Women, and the British School of Athens. However the main financial body throughout my study, as wellas the person who supported me emotionally, was my mother to whom I am most grateful. I will always be grateful to my dear friend Violetta who has been a sister to me all these years. I am grateful to my doctoral supervisor, Paul Buckland for teaching me the skills of research, guiding and inspiring me all the way, helping me with every problem, and editing the work. Thanks also to Joan, Helen, Pete and Tessa and Phil for sorting out computing frustrations and panics.

I would also like to thank Anaya Sarpaki who suggested the Santorini material, guided, advised, helped and inspired me and was always encouraging. To Peggy Sotirakopoulou I will always be obliged for advice and information about Akrotiri during the early days of the work. I am grateful to Professor Doumas for making available for study the Akrotiri material, as well as giving permission to use illustrations and photographs for the thesis. Also, many thanks are due to J.-P. Wild, P. Warren and C. Broodbank for their useful comments on silk.

I am obliged to Peter Skidmore for identifying the dipterous remains from Akrotiri. I would also like to thank Barry Kemp and the Egypt Exploration Society for providing the material from Amarna, Delwen Samuels for being so helpful, Marijke van der Veen for the Mons Claudianus material, Piotr Bienkowski for being so obliging during my visit to the Liverpool Museum, Simone Riehl for providing the Troy material, and John Wade for giving me the program for classical Greek and for reading through the relevant chapters.

Corrections and additions to the Egyptian part of the research was made possible by an award from the Leverhulme Trust, to whom due acknowledgement is made.

List of Tables

List of Figures

List of Plates

1. Introduction

Although the earliest research upon insects from archaeological sites dates back almost a century and a half, archaeoentomology, the study of insect fossils from archaeological contexts, is a relatively new field of research, still in its infancy in the Mediterranean region. Although there are a number of earlier studies of material from Egyptian mummies (e.g. Hope 1842), the first attempt to reconstruct a past environment from fossil insects from archaeological contexts was undertaken by Rondani, and was included in the report on the *Terramare* sites of the Parmense, south of the river Po (Strobel & Pigorini 1864). Rondani showed the possibility of reconstructing ancient environments based on entomological evidence. Since then, insects of the order Coleoptera (beetles) have proved very effective in the study of climatic and environmental change (Buckland & Coope 1991). In an archaeological context, the group can be further used to interpret a variety of facets of everyday life that would otherwise pass unnoticed in the record, such as aspects of economy, living conditions, hygiene and trade.

The immense scope and power of this research approach are perhaps best known from the pioneering studies in Britain of interglacial, glacial, Lateglacial, Holocene and urban archaeological sediments (e.g. Buckland & Coope 1991; Coope 1994; 1995; Hall & Kenward 1990; Kenward & Hall 1995; Buckland 1990). Work in the Mediterranean, however, has not been systematic, and has been carried out mostly by entomologists with no extensive knowledge of the archaeology involved (e.g. Alfieri 1931; dal Monte 1956), or archaeo-botanists with no deep under-standing of the entomological data. For the first time, at the site of Akrotiri on Santorini in the Greek Aegean, a systematic study of insect remains from an archaeological site has been undertaken in the region. However, a variety of problems have had to be confronted. Methodological difficulties occur as a result of the different conditions of preservation compared with North European sites. These rule out the use of the standard technique of paraffin (kerosene) flotation (Coope & Osborne 1968), and other methods of extracting the insect remains have had to be investigated.

The amazing preservation of archaeological material at the site of Akrotiri gave a unique opportunity to reconstruct in detail an environment over 3000 years ago, and this formed the starting point for research. Archaeobotanical research had already been undertaken on the site by Sarpaki (1987). A first examination of the material verified the preservation of insects (Panagiotakopulu 1989), and a detailed study of the insect faunas was therefore begun.

The main objectives of the study have been :

i) to consider the problems encountered because of the different preservation in Mediterranean sites with respect to palaeoentomology.

ii) to test the methodology already applied in Northern European sites, and, if necessary, to develop a new methodology.

iii) to evaluate the potential of insect remains in Mediterranean archaeo-logical sites.

iv) to investigate aspects of life, hygiene, living conditions, and economy, based on the insect fauna.

v) to investigate the existence of insect technologies, such as production of wax, dyes, etc.

vi) to investigate the past distribution of the synanthropic insect fauna.

vii) to consider the origins of the fauna.

viii) to examine the role of trade in relation to the archaeoentomo-logical evidence around the Eastern Mediterranean basin.

2. Methodology

Introduction

In the living world, insects are usually the most numerous animals, both in terms of taxonomic diversity and number of individuals. Recovery of their remains from archaeological contexts, however, is beset by a number of taphonomic problems. In this chapter, the losses of archaeo-entomological information on sites in the Mediterranean region as a result of taxonomic, taphonomic, sampling and processing methods are examined. By way of example, the difficulties during sampling and processing of material from Akrotiri are also discussed.

Unlike many North European countries, Greece does not have a long tradition of entomological research, and there is no complete comparative collection of Coleoptera in Greece. There are also no updated identification keys for much of the beetle fauna, although some groups, for example, the Chrysomelidae (*pars.*) (Gruev 1990), have been revised. In addition there is no checklist; the nearest ones, for Egypt (Alfieri 1976) and a partial one for Cyprus (Georghiou 1977) belong to different biogeographic provinces. The first attempt at a complete insect checklist for the complete Italian fauna was begun in 1991 by 240 specialists working for the Ministry of Environment and the Scientific Committee for the *Fauna d' Italia*, and this is still in progress (Reavey *et al.* 1994). As a result, the identification of Coleoptera is a major problem both for living and fossil communities, although much of the fauna is covered in *Die Käfer Mitteleuropas* (Freude, Harde & Lohse 1963-89). As well as the Coleoptera, larval Diptera, maggots, are also often found in archaeological deposits, as the puparia, the hardened skin of the last instar, inside which pupation takes place, but identification is often almost impossible because of a lack of information on the modern fauna, and the larval stages of many species have yet to be described (*cf.* Skidmore 1996).

Information loss

In palaeoecology, taphonomy is the process of changes of the organic remains after death (Behrensmeyer & Kidwell 1985). There are different taphonomic levels: biocoenosis (Fagerstrom 1964: 1198) concerns the living assemblage (*bio* in Greek means life, *enosis* union, combination, mixture), thanatocoenosis, the process of accumulation of the death assemblage (*thanatos* is death), taphocoenoesis meaning the post burial processes (*taphos* : grave) and finally oryctocoenosis, the fossil assemblage (*orycto* means mineral, excavated from the ground). Only the words biocoenosis and thanatocoenosis have entered the more general literature (*cf.* Brouwer 1967). Gilbert and Singer (1982) suggest that during palaeoecological research there are the following types of information losses: biotic, thanatic, taphic and sullegic. On most Mediterranean archaeological sites, the problem of losses can be immense. Waterlogged contexts are virtually unknown, and preservation usually relies upon desiccation or charring.

Thanatic

Fossil insects are usually preserved only under very dry or very wet conditions and insects living in other environments may not be preserved at all. Depending on the site, there are different ways of preservation of fossils:

1. Preservation in very wet conditions, waterlogged environments such as peats, lakes, wells, etc. This is the usual mode of preservation in northern sites (Buckland & Coope 1991).

2. Preservation by charring is also encountered, and may provide the only remains of an insect assemblage (e.g. Buckland 1981). Carbonized seeds and associated pests may be a result of human activity patterns or natural causes, such as the volcanic eruption in the case of Akrotiri. Preservation by charring is controlled by the duration of heating, the temperature, oxygen and moisture levels. As a result of these factors, the degree of charring can differ from one context to another.

3. Preservation by desiccation is characteristic of sites in very dry zones and the fossils are found intact.

4. Preservation by mineralization occurs as a replacement of the individual by inorganic substances. In the case of Akrotiri, the inorganic substances would appear to be related to faecal and other material (Sarpaki 1987). Sometimes mineralization makes the identification of the fossil a very difficult task as the cuticle can be covered completely by a thick layer of calcite and no diagnostic features of the animal are immediately detectable, although a cast may indicate the necessary diagnostic features.

Taphic

Possibilities of contamination and disturbance of deposits should always be considered. These include human activities as well as the activity of mammals, particularly rodents, other arthropods, and earthworms. It is essential that insect fossils are examined in combination with other available archaeological and palaeo-ecological information. This reduces the possibility of contaminants passing unnoticed. Bruchids, for example, in charred infested pulses could be considered as contaminants by palaeobotanists. Sometimes the quality of preservation makes the distinction between modern and fossil faunas extremely difficult. In the case of the sites examined as part of this thesis, Akrotiri, Amarna, Mons Claudianus, and Kahun, insect faunas were always examined with respect to the contexts in which they were found, and were considered for problems of possible modern contaminants.

Sullegic

Another point that should be made is the loss of information due to the lack of sampling for biological remains on many Mediterranean and Near Eastern sites. The quality of preservation of archaeological finds on several key sites (e.g. Amarna in Egypt (Kemp 1984; 1985), Çatal Hujuk, Turkey (Melaart 1962), and Akrotiri, Santorini, Greece (Doumas 1983)) is excellent. However, in the past, environmental research on such sites has been always considered secondary, if at all. In addition, the sampling methods employed were another reason for losses of information. Sampling with respect to the particularities of the site, securely dated contexts, as well as standard sizes of samples where applicable, are essential. A specialist on site is required when sampling for biological remains, since excavators rarely have the background to make decisions on sampling for palaeoecological research.

A sampling framework is also essential, and Sadler (1991) has discussed sampling for palaeoentomology from north European archaeological excavations in some detail. Mediterranean sites, however, are usually different from north European sites in several respects. The principal aims are the same: reconstruction of the economic life of a settlement and details of living conditions; these are areas where insect remains can make a significant contribution.

The preservation of insects can bring a completely different dimension to the choice of field methods and sampling decisions, and the whole approach of the specialist will be dictated by preservational parameters. Storerooms of a house, for example, should be sampled in detail, while bulk samples from floors on a site like Akrotiri, where the basic processing method is dry sieving, could mean a huge amount of work with very little results.

Laboratory Techniques

Since the late 1960s, the standard method for the recovery of insect fossils has been paraffin (kerosene) flotation (Coope & Osborne 1968). The technique is based on the wet sieving process, but it has been modified to include flotation with paraffin (kerosene), after disaggregation over a 300μm sieve, so as to recover the insect exoskeletal fragments. Several suggestions aiming to increase fossil recovery during paraffin flotation have been made (*cf.* Kenward 1974; Kenward *et al.* 1980; 1985). Although the method has been found to be ideal for waterlogged environments, it cannot be applied in all situations.

Experiments were conducted at Akrotiri. Paraffin flotation was found to be inappropriate because of the character of the samples, a sand-sized tephra, and the high temperatures during summer. The water, during the flotation session, heated up very quickly and, as a result, literally all the organic part of the sample floated. Wet sieving through a 300μm sieve was also attempted. However charred fossil insects tend to be very fragile, and only fragments of the cuticle were occasionally retrieved. The method finally adopted was dry sieving through 100μm and 300μm sieves, stacked one on the top of the other. In this way, the samples were separated into two different portions, according to size, which were then sorted under the microscope. Dry sieving is a very time consuming method, but it reduces the degree of destruction of carbonized fossils due to mechanical and water action.

Desiccated and mineralized specimens can be retrieved by the standard method for seeds, wet sieving, as the exoskeletons of desiccated fossils are quite resistant. Paraffin flotation would be theoretically ideal for these cases, but, as already mentioned, the high temperatures on site preclude its use. In fact, in the absence of systematic archaeo-entomological work, the only material retrieved was found during sorting for seeds.

Akrotiri

Sampling methods

The volcanic eruption that completely covered Akrotiri resulted in superb preservation (Fig. 2.1). Everything was found *in situ*: whole buildings, storerooms with pots, crops and flour. An archaeobotanical study was undertaken by Sarpaki (1987). This gives detailed information on the botanical remains from the West House. On the basis of this, the site offered an ideal opportunity for archaeo-entomological study, although the West House was already completely excavated when the above study was undertaken. All the samples from the House were kept in the storeroom of the excavation. These were mainly the contents of pots or from archaeological contexts deemed to be interesting.

During the process of the excavation, grab samples were taken from floors and the contents of pots were also sampled. Due to the difficulties of dry sieving, the size of grab samples had to be standardized to three litres instead of the five litres, which most previous archaeoentomological work had relied upon (*cf.* Sadler 1991). Three litres was also the maximum quantity studied from the contents of the pots, pithoi.

The House of the Ladies, Xeste 3, Xeste 4, The Delta Building and Pessos 17 were sampled so as to obtain a better idea of what life in the settlement was like from the insect point of view. Priority was given to contents of pots as well as samples from sewers. In total 267 samples were taken (including the samples studied from Αποθήκη 40).

Laboratory methods

Processing of the samples kept in Αποθήκη 40, mainly the contents of pots, was by dry sieving and the residues sorted under the microscope. The smaller ones, some of them under 50g, were directly sorted under the microscope. In order to speed up the sorting of grab samples from floors, paraffin flotation was attempted, as dry sieving seemed an unrealistic choice. A series of experiments was conducted using paraffin flotation as well as simply wet sieving. Eighteen soil samples were floated with paraffin, after having added counted specimens of modern charred insects. The same number of soil samples with added insects was processed by wet sieving. Almost no insects were retrieved in either experiment.

The poor success rate of the techniques led to the decision to continue using dry sieving till a new method appropriate for charred remains is developed. However, a further decision concerning samples from contexts other than pot contents had to be taken. After the wet sieving results, if the samples already sorted for biological remains by Dr. Sarpaki's team on site were found to be completely empty, these were left, while priority was given to archaeologically interesting samples.

The archaeoentomological samples were sorted using a binocular stereoscopic microscope at the excavation and at the Department of Archaeology and Prehistory, University of Sheffield. The modern comparative material was from collections in Doncaster Museum, Sheffield, and from the Natural History Museum in London. The identification of the insects was problematic and time consuming because of the lack of facilities as well as theabsence of a check list of the Greek fauna or a Mediterranean collection in the University.

Conclusion

Basic entomological background work, a complete insect checklist for the area as well as a competent identification key are absolutely necessary for the Mediterranean and Near Eastern areas. However, preliminary archaeo-entomological assessments of various sites and experimental work on taphonomic processes does give a better idea of the preservation of insects in different Mediterranean archaeological contexts.

Sampling frameworks in respect of the particularities of the sites excavated should be employed. The above demands the presence of a specialist during the excavation, as well as close collaboration with the excavator and specialists of different disciplines involved in the same project.

There is a need for experimental work so as to develop a new method for separating carbonized remains that will not be as time consuming as dry sieving. Archaeoentomological study, for the moment completely neglected, should be undertaken in excavations around the Mediterranean. The information that could be gained is unique and highly important, establishing not only the archaeological and biogeographical record of insects, but also their impact upon past agricultural systems and the ancient economy.

Fig. 2. 1. Preservation at Akrotiri on Thera : a complete pot with charred contents and (right) remains of a basket.

3. The Archaeological Record of Pests

Introduction

Whenever Man started storing foodstuffs and skins, insects came into the storerooms, either to exploit the artificially warm microclimate or to live directly on the stored material. Many species were already living in environments, which, in terms of microclimate and microenvironment, were very similar, if not identical to the synanthropic ones. These included the nests of birds and rodents, or under thick leaf litter. Several species were thus easily able to adapt to the new conditions, becoming residents in storerooms and places occupied by Man, and became specialised feeders on stored products.

Insects that infest contemporary storerooms have been a problem for agriculture from early prehistory (Buckland 1981) (Fig. 3.1). According to Dobie and others (1984), among the one hundred important food pests, eleven species of beetle (Coleoptera) and two of moth (Lepidoptera) are the most common (Table 3.1). These pests are the most destructive, as they multiply rapidly, adapt easily to a variety of ecological and climatic conditions and penetrate without difficulty into new stores. Not surprisingly, all of them have been recorded in archaeological contexts, except for *Callosobruchus maculatus* (F.). Either *C. maculatus* was not of such importance around the Mediterranean during early prehistory or, more probably, it has yet to be found. Only further research can answer the question.

Species	Size mm	Commodities infested	Place of origin	Archaeo-logical finds
a. Beetles (Coleoptera)				
Trogoderma granarium (Everts)	2-3	oilseeds, cereals	India	-
Oryzaephilus surinamensis (L.)	2.5-3.0	cereals etc.	?	+
Cryptolestes ferrugineus (Steph.)	2.5	most products	?	+
Rhyzopertha dominica (F.)	2-3	cereals	India	+
Lasioderma serricorne (F.)	2-3	processed foods		+
Tribolium castaneum (Hbst.)	3-4	cereals etc.	India	+
T. confusum (Jacq. du Val)	2-3	cereals etc.	Ethiopia	-
Callosobruchus maculatus (F.)	3-4	pulses	Tropical Africa	+
Sitophilus granarius (L.)	3-4	cereals	Near East?	+
S. oryzae (L.)	2.5-4.0	cereals	India?	+
S. zeamais Mots.	3-4	cereals	India?	+
b. Moths (Lepidoptera)				
Ephestia cautella (Walk)	1.5-2.0	cereals etc.	Tropics	-
E. kuehniella (Zell.)	1.5-2.5	cereals, nuts	?	-

Table 3 -1 Principal pests in storerooms (revised from Dobie *et al.* 1984)

The need to transport grain and other food commodities, initially on a local and then on a more regional scale, has lead to insect pests being easily spread around the world and many are now cosmopolitan (Munro 1966). However, different insect pests have different food preferences and different temperature requirements. These factors define abundance in various parts of the world. For example, *Trogoderma granarium* (Everts) is a major pest of stored grain in India, and has also been recorded as destructive in Europe, but it has only lately become established in the south western USA, and is not yet established in Canada, despite being introduced several times through shipments (Becker 1978).

The following section summarises the available habitat data and the fossil record for the major insect pests of stored products.

The archaeological records

The sap beetles (Nitidulidae)

Certain species of the family of Nitidulidae, such as *Meligethes aeneus*, a pollen beetle, can also be found as pests of crops (Harde 1984). Other species, including *Omosita* spp. and *Nitidula* spp., are associated with fungus, carrion, old bones, and old skins (Hinton 1945; Koch 1989), and are therefore occasionally synanthropic. *Omosita discoides* (F.) and *O. colon* (L.) have been recorded from several sites including a Neolithic site in England, at Runnymede, Surrey (Robinson 1991), and *Nitidula bipunctata* (L.) occurs in sediments from the Bronze Age well at Wilsford in Wiltshire (Osborne 1969); all are likely to be native to Europe.

***Necrobia* spp.** (Cleridae)

Necrobia rufipes (Deg.) feeds on carrion during the fermentation stage after the carcass has dried out. It has also been recovered from fish, bones, cheese and ham (Dillon & Dillon 1972). The insect is widely known from Egyptian mummies (Levinson & Levinson 1994), including that of Rameses II (1224 BC) from Thebes (Alluaud 1908). One of the earliest identifications of a fossil insect in an archaeological context is of this species from a Ptolemaic mummy by Hope (1834). More recently, the Manchester Museum Mummy Project, involving autopsies on a number of Egyptian mummies, noted the insect in the wrappings of mummy 1770, dated to AD 323-440 (Curry 1979, 115). The only European fossil comes from late medieval deposits at Neuss in the Rhineland (Koch 1971), and the species would appear to be a late introduction. Its congener, *N. violacea* (L.) was also recorded by Hope (1834) from a Ptolemaic mummy, and it also does not seem to reach northern Europe until the medieval period (Kenward & Hall 1995). The absence of *Necrobia* spp from Roman sites is surprising, and Buckland (1997) in reviewing the Anglo-Scandinavian record of *N. violacea*, *Dermestes lardarius* L. and *Attagenus pellio* (L.) from York, has suggested that this may reflect differences in meat processing, if the reasons are not taphonomic.

The skin and hide beetles (Dermestidae)

The genus *Dermestes* includes a number of species whose larvae and adults live on carcasses, bones and carrion. *D. frischi* Kug. and *D. ater* Deg. have also been recorded from dried fish, dead locusts, dry rabbit and other skins (Peacock 1993). Able to complete their life cycles in only five weeks (Hinton 1945), the species have been found in cacao warehouses in Britain and Germany, in granaries, mills and houses, feeding on dead insects. They are also known to infest silkworm cocoons, dried sponges, fungi and dog biscuits (Hinton 1945). *D. frischi* is not a household pest and is normally connected with animal products (Peacock 1993). It is recorded from the mummy of Rameses II, who died in 1235 BC (Alluaud 1908), and from Pum II of 2600 BC (Cockburn *et al.* 1975) in Egypt. Also, *D. frischi* and *D. ater* were found in the mummy of Han-Em-Ken-Esi, a tomb builder, and priest, from *ca.* 1000 BC (Strong 1981). *D. frischi* specimens were also recorded earlier from the heads and canopic jars of Egyptian mummies (Hope 1834; Lesne 1930).

The origins of *Dermestes* are uncertain, but *D. ater* is known from a Han tomb of 100 BC at Ma-wang-tui, Changsha, Hunan Province, China (Chu & Wang 1975), where it was recovered together with another Dermestid *Trogoderma persicum* Pic. and the rice weevil, *Sitophilus oryzae* (L.). Another species of *Dermestes*, *D. lanarius* Ill., was recovered from a Late Bronze Age well at Wilsford, Wiltshire, England (Osborne 1969). The beetle is not wholly synanthropic (Minoranskiy 1969), and has failed to become cosmopolitan. Aitken (1975) noted only one accidental import to Britain, where it is no longer native, in potatoes from Greece. Robinson (1991) has recorded *D. lardarius* Deg. from a Bronze Age site at Runnymede on the Thames, and Kenward and Hall (1995) note it from Anglo-Scandinavian York. The older records from mummies, however, may need to be treated with some circumspection, in that infestation may have occurred during transport and storage in museums.

Thorictodes heydeni Reitt. was first described from fossil material as *Thaumaphrastus karanensis* Blais. by Blaisdell (1927) from a plant gall from Roman Karanis in Egypt. Van Emden (1951) later noted the species from a wide range of stored products and indicated that it had a wide distribution in synanthropic situations in the Tropics, with occasional records from temperate regions.

Specimens of *Anthrenus* sp., a genus sometimes associated with stored products (Peacock 1993), were recovered from Roman Mons Claudianus in the Egyptian desert (Panagiotakopulu & van der Veen 1997). Several species of Dermestidae, *Attagenus* spp., *Ctesias* spp. and *Thylodrias contractus* Motsch, have been identified from the garbage dumps of Sahr-i Sokhta in Mesopotamia, from contexts dated to 2700 to 2200 BC (Costantini *et al.* 1977). The above species feed on other insects and organic material, and they can be found together with Man or in the wild (Peacock 1993). Perhaps their abundance at early Sahr-i Sokhta points to a Near Eastern origin for the beetles.

Trogoderma granarium (Everts) or the khapra beetle is a dermestid, now cosmopolitan, that prefers dried vegetable matter to animal matter as food, but is able to feed on either. Its larvae are extremely resistant to starvation, and under favourable conditions it multiplies very rapidly. The larvae nest in crevices of buildings, bins, etc., which are difficult to reach with insecticides. The beetle probably originated in India and it has spread around the world as a result of accidental transport by Man (Munro 1966; Hunter *et al.* 1973). There are no fossil records, although the record of *T. persicum* from first century BC China (Chu & Wang 1975) hints at an Asiatic origin for the genus.

The cadelle (*Tenebrioides mauritanicus* L. (Ostomidae))

The ostomid *Tenebrioides mauritanicus* L., the cadelle, is a very well known pest of flour mills that becomes established in the woodwork of storage warehouses, and invades cereal products during storage or shipment. It feeds on grain, flour, seeds and other insects (Dillon & Dillon 1972), but is often more destructive of sacking than flour (Munro 1966). Its earliest records are from Roman York (Hall *et al.* 1980; Kenward & Williams 1979), although its origins are likely to be Mediterranean.

The saw toothed grain beetles (*Oryzaephilus* spp. (Silvanidae))

Oryzaephilus surinamensis (L.), and *O. mercator* (F.), the saw toothed grain beetles, are cosmopolitan pests of stored grain and grain products. *O. surinamensis* is a secondary pest, feeding on grain already attacked by primary pests, especially the grain weevil, *Sitophilus granarius*. It is found in warehouses, mills, farmhouses with granaries, and brewery silos (Zacher 1927). The adults are also carnivorous, living on the young larvae and caterpillars of other pests in the grain and sometimes they will even consume their own species (Zacher 1927). The beetle is able to overwinter in unheated stores (Solomon & Adamson 1955), but it cannot complete its development at temperatures below 18^0 C, and does not thrive below 22^0 C (Howe 1965a). Its distribution around the world is mostly a result of trade, although its primary *pabulum* is likely to have been under loose bark (Hunter *et al.* 1973). Birds' nests would have been a suitable niche for the saw toothed grain beetle before it became a storeroom pest. Its earliest record in the Mediterranean comes from Neolithic Mandalo in Macedonia (Valamoti & Buckland 1995). It has also been recorded from Tut'ankhamun's tomb, 1345 BC, in Egypt (Alfieri 1931), and from Late Bronze Age Akrotiri on Santorini (Panagiotakopulu & Buckland 1991). *O. surinamensis* is widespread in Roman deposits in the western part of the Empire (Buckland 1981; Pals & Hakbijl 1992; Yvinec 1997), but there is currently no fossil record for *O. mercator*.

The flat grain beetles (*Cryptolestes* spp. (Cucujidae))

Cryptolestes ferrugineus (Steph.) is a secondary pest of cereals, nuts, milled products, dried fruits, and related products. It is not a primary pest, but a mould and detritus feeder (Halstead 1993). *C. ferrugineus* has been found in natural habitats in England, under bark (Donisthorpe 1939), where it is probably native and able to overwinter (Solomon & Adamson 1955). Fossil records are known from England, where it was recovered from the first century AD harbour sediments at Fishbourne, Sussex, England (Osborne 1971b), from first and fourth century deposits in York (Kenward & Williams 1979; Buckland 1976a) and from a fourth century burnt grain deposit at Droitwich, Worcestershire (Osborne 1977). There are also Roman records from Laurium in the Netherlands (Pals & Hakbijl 1992) and Amiens, France (Yvinec 1997).

Cryptolestes turcicus (Grouv.) is associated with stored products (Hodge & Jones 1995), and it has been recorded from flour mills (Munro 1966). The species appears to thrive in warmer climates, and it is not surprising that its first record comes from the first century AD Roman fort at Mons Claudianus in Egypt (Panagiotakopulu & van der Veen 1997).

The lesser grain borer (*Rhyzopertha dominica* (F.) (Bostrychidae))

The lesser grain borer, *Rhyzopertha dominica* (F.), although one of the smallest beetles harmful to grain, can cause serious damage to stored crops. It is a usual pest on grain in warmer countries (Munro 1966), and it is also recorded from a wide variety of crops such as wheat, barley, millet, rice, maize, and sorghum, as well as dried potatoes, manioc roots, waternuts and biscuits (Hickin 1963). It was originally described from South America (Munro 1966), but may have originated in India (Kislev 1991), and it has been carried through commerce around the world. *R. dominica* has been recovered, from botanical remains that have been dated to 1900-1800 BC, from Kahun in Egypt, in a vessel from Tut'ankhamun's tomb of 1345 BC (Alfieri 1931), from Late Bronze age Akrotiri (Panagiotakopulu & Buckland 1991) and Roman Mons Claudianus (Panagiotakopulu & van der Veen 1997).

The drug store-beetle (*Stegobium paniceum* (L.) (Anobiidae))

The drug-store beetle, *Stegobium paniceum* (L.), breeds in starchy materials, including cereals and many other commodities such as spices, cocoa beans and liquorice (Munro 1966); it may be abundant in pigeon's nests (Woodroffe 1953). Also known as the biscuit beetle, from its frequency in rations during the First World War (Osborne 1997), it is known from biscuits and bread, and was recovered from deposits of 1399 BC and bread of 2049 BC from Egypt (Chadwick & Leek 1972). At Dier el Medina, it is recorded from offerings in the tomb of Kha (XVIII Dynasty, ca. 3400 bp) (Levinson & Levinson 1994). It also occurs in material from the West House at Late Bronze Age Akrotiri (Panagiotakopulu & Buckland 1991), and is known from sites from the Late Bronze Age onwards in England (e.g. Osborne 1969; Robinson 1991).

The cigarette beetle (*Lasioderma serricorne* (F.) (Anobiidae))

Lasioderma serricorne (F.), the cigarette beetle, is largely a pest of stored tobacco. However, it may also damage a great variety of dried, stored materials of plant and animal origin (White 1990). It has been recorded in dried vegetable products, for example, tobacco, cayenne pepper, coriander, caraway, yeast, dried figs, etc. (Dillon & Dillon 1972; Munro 1966). Its origin could be also Near Eastern, and its first archaeological record is from the Late Bronze Age site at Akrotiri (Panagiotakopulu & Buckland 1991). It has also been found in Tut'ankhamun's tomb in Egypt (Alfieri 1931; Zacher 1937).

The spider beetles (Ptinidae)

Gibbium psylloides Cz. and *G. aequinoctiale* Boield. are flightless beetles, found in houses, hotels, latrines and granaries. They feed on grain and a large variety of vegetable products such as bread, cake, spices, cotton, hay, wool, leather, etc. (Dillon & Dillon 1972), but are also frequent in human faeces in dark places, including mines, where the ambient temperature is up to 33°C (Constantine 1995).

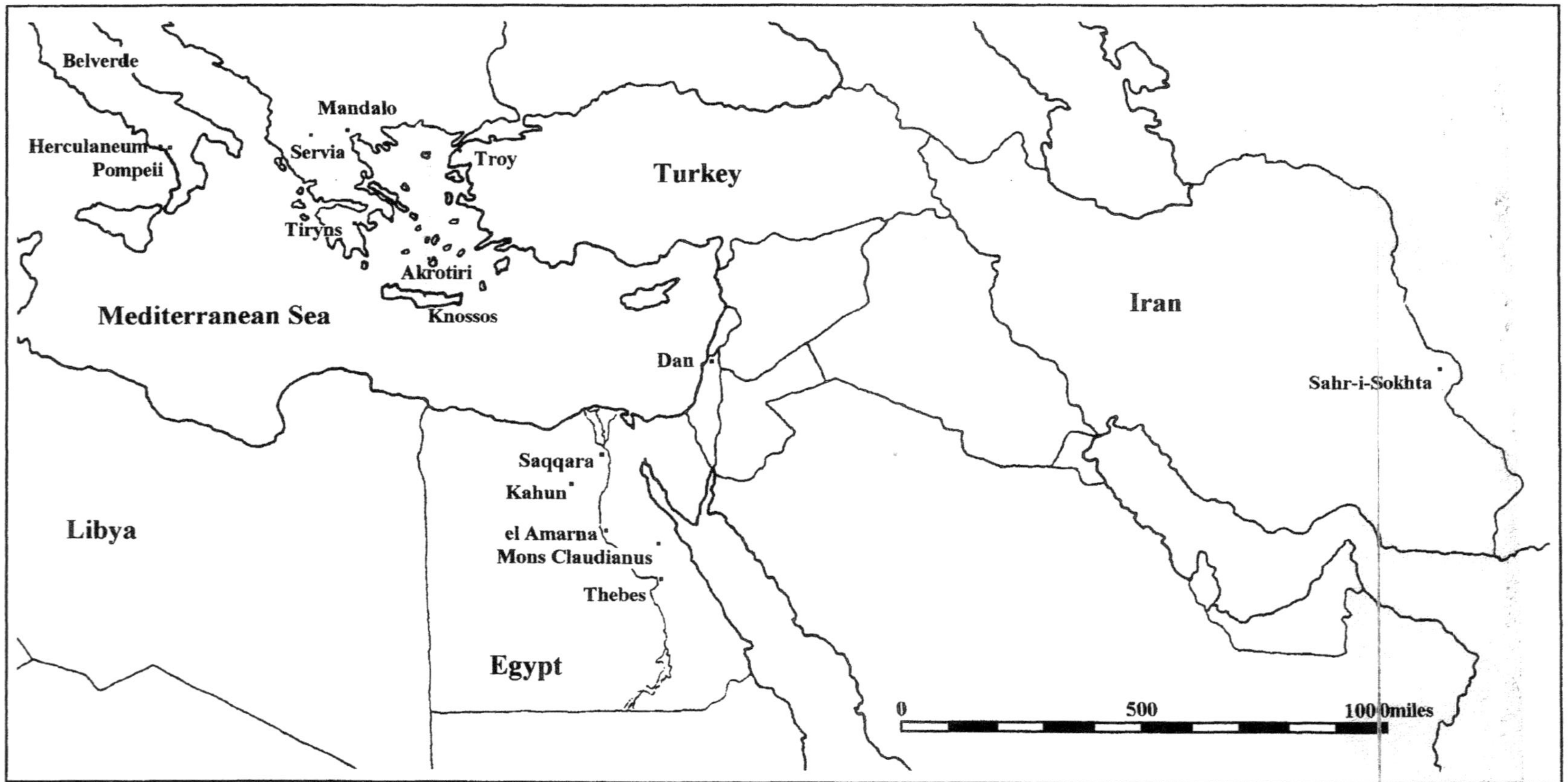

Fig. 3.1 Location map of Mediterranean and Near Eastern sites with evidence of pests of stored products.

The two species are very difficult to distinguish, and all the modern British records of *Gibbium* are now known to be of *G. aequinoctiale* (Hodge & Jones 1995). The preference for high temperatures (Constantine 1995) suggests a Mediterranean to tropical origin, and the earliest records came from different contexts at Sahr-i Sokhta in Mesopotamia, as early as 2400 BC (Costantini *et al.* 1977). The specimens were identified as *G. psylloides,* but it is more probable that they are of *G. aequinoctiale.* During the course of this research, *Gibbium* sp. was also retrieved from Roman Mons Claudianus (Panagiotakopulu & van der Veen 1997), and *G. aequinoctiale* occurs in large numbers in samples from Pharaonic Amarna (Panagiotakopulu, in prep.).

Niptus hololeucus (Fald.), the golden spider beetle, is now a nearly cosmopolitan pest, although absent from the Tropics. It is a long established pest in stores and houses and can be also found in the nests of bees, wasps and birds (Solomon & Adamson 1955). Its primary habitat may be birds' nests (Koch 1971), and its larvae may live off a range of products, including corn, semolina, oats and other starch-rich foodstuffs, as well as rotten wood. The adults appear to prefer textiles, although they can be also found in a great range of foodstuffs (Zacher 1927), and may feed off the debris of other insects. As this beetle is flightless, it is exclusively spread by Man (Lindroth *et al.* 1973). Only becoming widespread since the middle of the last century, the fossil record includes fourth century York (Buckland 1976b), medieval London (Tyers 1988), and fifteenth century Neuss in the Rhineland (Koch 1971).

Tipnus unicolor (Pill.) is a flightless beetle, whose primary habitat appears to be the nests of social insects, mammals and birds (Horion 1961). It is also found in barns and stables, in hay and straw waste (Koch 1971), and in cereal debris (Howe 1955). In the wild, it is recorded from birds' nests, and rotten trees (Palm 1959), and also in marmot and mouse burrows and bumblebee nests (Harde 1984). The earliest records appear to be from Roman Britain, where it appears to have become widespread in synanthropic situations (e.g. Hall & Kenward 1990; Girling 1989). A possible pre-Roman record from Thorne Moors, south Yorkshire (Roper 1996), has recently been revised to *Ptinus* sp. (Roper, pers. comm.).

***Aglenus brunneus* (Gyll.) (Colydiidae)**

Aglenus brunneus (Gyll.) is a flightless blind insect which probably originally lived under leaf litter in the forest. It is found in manure heaps, tan-pits, refuse of corn bins, old props in cellars, hotbeds, etc. (Fowler 1889). In Germany, it has been recorded in dark places such as cellars, stables and tanpits, under manure in greenhouses, under mouldy straw, among sacks of old bones, in haystacks, in cowsheds, and in mill refuse. In California, it has also been found destroying mushrooms in commercial beds (Hinton 1945). The beetle can be found in very old grain residues (Woodroffe 1967), and now appears to be totally dependent upon Man for its distribution (Horion 1961). Its British archaeological record has been discussed by Kenward (1975; 1976), and its existence in archaeological contexts should imply human activities and trade. In Egypt, it is recorded from the fort at the Roman quarry at Mons Claudianus (Panagiotakopulu & van der Veen 1997). In Britain, a record from Neolithic levels at the Sweet Track, Somerset (Girling 1984) is probably from its primary habitat, forest litter, but finds from Iron Age Farmoor, Oxfordshire (Robinson 1979), Roman York, (Kenward & Williams 1979; Hall *et al.* 1980), and later sites (Kenward 1975; 1976) are likely to be anthropochorous.

The flour beetles (*Tribolium* spp. (Tenebrionidae))

Tribolium castaneum (Hbst.) is a secondary pest of grain and an important pest on cereal products. It is not cold hardy, but may become permanently established in well heated buildings, and it has become cosmopolitan (Brendell 1975). It can be found in the wild in India, wherefrom it probably originated (Hinton 1948). It is also recorded from North America, and North Africa, where it has also been found under bark (Horion 1956). *Tribolium confusum* (Jacq.) is a cosmopolitan pest. It got its name because, for many years, it was confused with *T. castaneum,* to which it is very similar. It is a general feeder on farinaceous foods, more cold resistant than *T. castaneum*, and is undoubtedly the more important insect pest of prepared cereal foods, found in granaries, mills and warehouses

(Brendell 1975). Two other species of the genus *Tribolium, T. madens* (Charp.) and *T. destructor* (Uytt.), are both pests in flour mills and of farm stored grain (Brendell 1975). Brendell (*op. cit.*) agrees with Hinton (1948) that *T. confusum* and *T. destructor* are of African origin, and that *T. castaneum* is Indian. Kaszab (1969), however, believes that *T. destructor* is of South American origin, and places it in the sub-genus *Stene* with *T. confusum*, so both should originate from the same zoogeographic area.

Alfieri (in Andres 1931) recorded *Tribolium* sp. from an Egyptian tomb of the mid third millennium BC, but it is uncertain whether the species was *T. castaneum* or *confusum*, although he later (1976) notes the latter species from an unlocalised offering in a pot dated to *ca.* 1000 bc. *T. castaneum* is recorded from Egypt 500 years earlier (Zacher 1937). It is also recorded from a pithos that contained flour in the Delta building at Akrotiri on Santorini (Panagiotakopulu & Buckland 1991), and Brendell (pers. comm.) has recently confirmed its presence at Amarna. It is also present in samples from the Roman fort at Mons Claudianus in Egypt (Panagiotakopulu & van der Veen 1997). *T. castaneum* was found at the fourth century Roman villa at Droitwich, Worcestershire, England, together with *O. surinamensis* and *C. ferrugineus* (Osborne 1977), and it also occurs in a late Roman well in York (Hall *et al.* 1980) and in a similarly dated well at Chichester in southern England (Girling 1989). Identifications to the generic level extend back to the VI Dynasty (*ca.* 2345-2181 bc) in Egypt (Andres 1931).

The small eyed flour beetles (*Palorus* spp. (Tenebrionidae))

Palorus ratzeburgi (Wiss.) is a pest in cereal products and other commodities, being found particularly in mouldy grain residues previously attacked by grain weevils (Brendell 1975). The small eyed flour beetle feeds predominantly on the faeces of *S. granarius*, and is probably a predator on other pests (Pals & Hakbijl 1992). The earliest find is from pig coprolites from the Workmen's village at Amarna, Egypt, of the thirteenth-fourteenth century BC (Panagiotakopulu 1999). Other fossil records came from a second century shipwreck at the Roman fort of Laurium in Woerden, Holland (Pals & Hakbijl 1992), second century Amiens (Yvinec 1997), and Roman deposits in Britain, including York (Kenward & Williams 1979). Osborne (1971b) records a congener, *P. subdepressus* Woll., a pest rarely imported into Britain today (Brendell 1975), from second century deposits in the Roman town at Alcester, Warwickshire.

The lesser mealworm beetle (*Alphitobius diaperinus* (Panz.) (Tenebrionidae))

Alphitobius diaperinus (Panz.) is an omnivorous feeder, associated with grains, flour, leather, bones, etc., which may also be cannibalistic (Brendell 1975). Its earliest fossil record comes from the Amarna (Panagiotakopulu, in prep.) and it is present in the fort associated with the Roman quarry at Mons Claudianus in Egypt (Panagiotakopulu & van der Veen 1997). By the Roman period, it had also reached York (Hall & Kenward 1990).

The mealworms (*Tenebrio* spp. (Tenebrionidae))

The mealworms, *Tenebrio molitor* (L.) and *T. obscurus* (F.) are nocturnal beetles and prefer dark places. The two species are scavengers and feed on decaying grain and cereals that have been milled and preferably are damp and mouldy. They also feed on biscuits, flour, meat scraps, feathers, the bodies of dead insects, etc. (Solomon & Adamson 1955; Hunter *et al.* 1973). The earliest record of *T. molitor* is a charred specimen from deposits of carbonised seeds from the site of Troy (Panagiotakopulu, in S. Riehl, in prep.), of *ca.* 2000 BC, in Asia Minor, Turkey. All the other records of *Tenebrio*, from first century York (Kenward & Williams 1979), second century Alcester, Warwickshire (Osborne 1971b), and later sites in Britain have proved to be of *T. obscurus* (Kenward & Allison 1994), although *T. molitor* does occur in the Netherlands at Roman Laurium (Pals & Hakbijl 1992), and a specimen only identifiable to the generic level is known from Roman Mons Claudianus in Egypt (Panagiotakopulu & van der Veen 1997).

The seed beetles or bean weevils (Bruchidae)

Most members of this family infest the seeds of legumes in the field. The eggs are laid on the young legume seed pods and the larvae bore into the developing seeds.

The life cycle is completed in the seed and adults emerge by biting their way out. Several larvae may feed on one seed (Southgate 1978). The adults fly readily and are common sitting on flowers and in legumes in springtime. At the present day, the most common species of the family Bruchidae in Southern Europe is *Bruchus rufipes* Hbst. (Harde 1984). Bruchids infesting *Pisum sativum* seeds were recorded in a site from the Ukraine, dated to 3500 BC (Pashkevich, in Kislev 1991, 132). Bruchid infestation is also mentioned in the report on the Early Neolithic site at Belverde in Italy (Oliva 1939). Bruchid larvae, not identifiable to the species level, have been found in deposits from Troy dated to 2000 BC (Panagiotakopulu, unpubl.). *Bruchus rufimanis* Bohe. is present from the Iron Age at least in Britain (Caseldine 1987), and *B. rufipes* was found infesting *Lathyrus clymenum* at Late Bronze Age Akrotiri, Santorini (Panagiotakopulu & Buckland 1991). *Bruchus* sp. was recorded infesting *Vicia ervilia* from 12th century BC Tiryns, Greece (Kroll 1982), and *Bruchidius* sp. was found infesting *Acacia* seeds, dated to 1900-1800 BC from Kahun in Egypt. Bruchids were noted infesting pulses from Tut'ankhamun's tomb (Vartavan 1990), and also an apparently undescribed species occurred in lentils, *Lens culinaris,* from an Egyptian deposit of 215 ± 48 BC (Burleigh & Southgate 1975). Bruchids were also found infesting *Vicia ervilia* seeds from 11th century BC Dan in Israel (Kislev 1991, 132).

Whilst *Bruchus rufipes* is the most common Bruchid field pest in the Continental Europe and the Mediterranean region (Harde 1984), several other species are known as fossils from Britain, including both *Bruchus atomarius* (L.) and *B. loti* Payk. from Neolithic Runnymede in the Thames valley (Robinson 1991). They are as likely to have been part of the natural fauna as to have been imported with seed for crops.

The grain weevils (*Sitophilus* spp. (Curculionidae))

Sitophilus granarius (L.), the grain weevil, is a synanthropic species, recorded from a wide range of stored products including wheat, rye, barley, maize, oats, buckwheat, millet, chickpeas and even chestnuts, acorns and cornmeal (Hoffman 1954). It is usually, however, found in stored cereal crops. In contrast with its congener, *S. oryzae* (L.), it has never been found breeding in the field. Despite the fact that it is flightless, it has become cosmopolitan, and one of the most important pests of stored grain around the world, able to overwinter in unheated granaries and therefore easily adjusting to colder climates, such as those of Northern Europe. Howe (1965b) has suggested that acorns might have been the primary host of the weevil, but it is more likely that *S. granarius* moved from the nests of rodents to the stores of man, even before the beginnings of settled agriculture (Buckland 1981).

Site	Country	Date	Crop
Smeerenburg	Norway	17th C AD	*Vicia faba*
Matdanits-koye	Ukraine	3500 BC	*Pisum sativum*
Oxford	England	12-13th C AD	*V. faba*
Belverde	Italy	Eneolithic	*V. faba*
Pompeii	Italy	79 AD	*V. faba*
Akrotiri	Greece	1500 BC	*Lens culinaris* *Lathyrus clymenum* *L. cicera/L.sativus*
Tiryns	Greece	12 C BC	*Vicia ervilia*
Dan	Israel	11th C BC	*V. ervilia*
Gritelle	Turkey	Crusader period	fava beans (N. Miller pers. comm.)
----------	Egypt	2nd C BC	*Lens culinaris*

Table 3 - 2 : List of infested pulses from archaeological sites (revised from Kislev 1991).

The earliest evidence of *S. granarius* comes from a site at Neolithic Servia in Macedonia (Hubbard 1979, 227), where casts of the insect are imprinted in a piece of pottery. *S. granarius* is also known from Middle Bronze Age Crete (Jones 1984), and the Late Bronze Age settlement of Akrotiri (Panagiotakopulu & Buckland 1991). Solomon (1965) records it from barley deposited in a tomb beneath the step pyramid of Saqqarah about 2300 BC, and Helbaek (in Solomon 1965) notes further specimens of *Sitophilus* 600 years older from another tomb in Saqqarah. The earliest record from Western Europe has recently come from Early Neolithic Germany, 5000-4000 BC (Büchner & Wolf 1997).

Sitophilus oryzae (L.) is also one of the most destructive pests of stored grain. In tropical regions, the adult weevils fly from granaries or other sources of grain to fields of corn, wheat and rice, and there start infestations that may completely destroy the grain during storage (Hill 1975). In Europe, it has become the most common of the imported pests of grain shipments. There are no published prehistoric or Classical records from Europe or Africa, and its first archaeological record comes from a Han tomb in China dated to 100 BC (Chu & Wang 1975). In Britain, its first record is from a 15th century cesspit in Southampton (Grove 1997).

The flour moths (Lepidoptera)

The Mediterranean flour moth, *Ephestia kuhniella* (Zell.) was formerly the most destructive pest on flour. The larvae spin threads wherever they go, webbing and matting the flour particles together. *E. cautella* (Walk.) is also very troublesome, and a common pest in seed stores (O' Farrell & Butler 1948). Unfortunately, the chances of such soft bodied animals being preserved in an identifiable state are slight.

The Angoumois moth (Lepidoptera)

Sitotroga cerealella, better known as the Angoumois grain moth - from the province of Angoumois in France, where it is known to have been injurious since 1736, is now cosmopolitan in distribution (Munro 1966). It flies to the fields of ripening corn and wheat and lays eggs on the ears of wheat or the corn kernels. The main damage happens from the harvest until the wheat is threshed and stored. After threshing, infestation is restricted to the surface grain as the soft bodied moths cannot penetrate the heaps of grain. It is possible that the species was a pest of wild grasses in South Africa, and that it was introduced to Europe (Joubert 1966), but, in the absence of a fossil record, this must remain uncertain.

Diptera in Archaeological Deposits

Several species of true fly, the Diptera, are also occasionally found in stored products and their residues, and fragments, particularly of the pupae and puparia are very frequently found in archaeological deposits. For example, the house fly, *Musca domestica* L., and the stable fly *Stomoxys calcitrans* (L.) have been recorded from several Roman and later sites in Northern Europe (Skidmore 1996), although there has been virtually no work in the Mediterranean. *Musca domestica* the house fly and examples of *Piophila* spp. were recovered in the mummy 1770 during the Manchester Museum Project (Curry 1979). *Piophila* spp. flies are more commonly known as the cheese skipper, and are domestic pests (Oldroyd 1964). Species are known to breed in cheese, bacon, ham and similar food-stuffs and may be eaten by accident, whereupon the maggots may cause bleeding of the intestine. (Zumpt 1965; Oldroyd 1964).

The excavation at Sahr-i Sohta in Meso-potamia provides the earliest records of synanthropic flies, Muscidae, and puparia of Cyclorrhapha, all of them found in the garbage deposits, dated to around 2600 BC (Costantini *et al.* 1977). Skidmore's recent thesis (1996) on fossil Diptera examines 43 assemblages from England (including Lindow Man II and III, Thorne Moor, and urban sites in Worcester and York), Wales, Iceland, and West Greenland, and shows the great value of dipterous fossils and the potential of their study.

Ants (Hymenoptera, Formicidae)

Hypoponera punctatissima (Rog.), an ant, which is known as a pest of heated buildings in northern Europe (Ólafsson & Richter 1985), was found in the material from Akrotiri, Greece, and is also known from the Roman sewer in York (Buckland

1976a), as well as post-medieval Iceland (Buckland *et al.* 1992). The presently more common synanthropic pest species, known as Pharoah's ant, *Monomorium pharaonis* (L.) has yet to be found fossil.

Hymenoptera, Parasitica

Bracon hebetor Say was recorded in a bread offering from 1399 BC from Egypt, together with *S. paniceum* upon which it is a parasite (Chaddick & Leek 1972).

Mites

The mites are very small arthropods related to spiders. A number, such as the flour mite, *Acarus siro* L., are frequent pests. Mites in general are very common fossils in archaeological deposits. but only recently has there been a concerted attempt at identification (Schelvis 1997). Material has been published from a number of sites, including Late Neolithic Kolhorn in the Netherlands (Schelvis 1990; 1992), and Viking (van de Bund 1988) and Roman York (Denford 1976).

Taxon	British Isles	Germany	Italy	Greece	Turkey	Israel	Egypt	China	USA
Coleoptera									
Carabidae									
Laemostenus terricola (Hbst.)	1st C AD[1]								
Cleridae									
Necrobia violacea (L.)	10-11th C AD[2]						Ptolemaic[3]		
N. rufipes Deg.		15th C AD[4]					ca. 1224 BC[5]		
Dermestidae									
Dermestes frischi Kug.							ca.2600 BC[5]		
D. carnivorus F.							Pharaonic?[6,7]		
D. laniarius Ill.	Bronze Age[8]								
D. ater Deg.							ca.1000 BC[9]	1st C AD[10]	
D. lardarius L.	Bronze Age[11]								ca. 1650 AD[12]
Dermestes sp.	Early Holocene[13]						Pharaonic[14]		
Trogoderma persicum Pic.								1st C AD[10]	
Attagenus pellio (L.)	4th C AD[15]								
Anthrenus verbasci L.		15th C AD[4]							
A. museorum (L.)		15th C AD[4]							
Anthrenus sp.	4th C AD[16]						Pharaonic[14]		
Thorictodes heydeni Reitt.							2nd C AD[18]		
Trogositidae									
Tenebrioides mauritanicus (L.)	1st C AD[1]								ca. 1650 AD[12]
Nitidulidae									
Carpophilus hemipterus (L.)									ca. 1650 AD[12]
Omosita colon (L.)	Bronze Age[11]								

Nitidula bipunctata (L.)	Bronze Age[8]	15 C AD[4]							
Cucujidae									
Oryzaephilus surinamensis (L.)	1st C AD[1]	1st C AD[4]		ca. 3000 BC[18]			ca. 1336 BC[19]		
Oryzaephilus sp.	1st C AD[1]		AD 79[20]	ca. 1500 BC[21]					
Cryptolestes turcicus Grouv.							2nd C AD[22]		
C. ferrugineus (Steph.)	1st C AD[1]						2nd C AD[22]		
Colydiidae									
Aglenus brunneus (Gyll.)	ca. 3200 BC[23]	15th C AD[4]							
Bostrychidae									
Rhizopertha dominica (F.)	15th C AD[24]			ca. 1500 BC[21]			ca. 1327 BC[19]		
Anobiidae									
Stegobium paniceum (L.)	Bronze Age[8]						ca. 3400 BC[25]		
Lasioderma serricorne (F.)							ca. 1327 BC[26]		
Ptinidae									
Gibbium aequinoctiale (Goez.)							ca. 1327 BC[26]		
Gibbium sp.							1550-1050 BC[27]		
Niptus hololeucus (Fald.)	4th C AD[28]	15th C AD[4]							
Tipnus unicolor (Pill. & Mitt.)	1st C AD[16]	15th C AD[4]							
Ptinus fur (L.)	Bronze Age[8]	1st C AD[29]							ca. 1650 AD[12]
Tenebrionidae									
Palorus subdepressus (Woll.)	2nd C AD[30]								
P. ratzeburgi (Wissm.)	1st C AD[1]						ca. 1350 BC[31]		
Tribolium castaneum (Hbst.)	4th C AD[15]						ca. 1327 BC[19]		
Tribolium sp.	2nd C AD[32]						2345-2181 BC[33]		
Alphitobius diaperinus (Panz.)	2nd C AD[32]						ca. 1350 BC[31]		
Gnathocerus cornutus (F.)									ca. 1650 AD[12]
Tenebrio opacus Duft.	1st C AD[34]								
T. molitor L.					ca. 2000 BC[35]				
Bruchidae									
Bruchus loti Payk.	5400-3500 BC[11]								
B. atomarius (L.)	Bronze Age[11]								
B. rufimanus Bohe.	Iron Age[36]								
B. rufipes (Hbst.)				ca. 1500 BC[21]					

Curculionidae									
Sitophilus granarius (L.)	1st C AD[37]	ca. 5000 C[38]	AD 79[20]	ca. 3000 BC[39]		9th-7th C BC[40]	2630-2611 BC[41]		ca. 1650 AD[12]
S. oryzae (L.)								1st C AD[10]	ca. 1650 AD[12]

Key

1 York (Kenward & Williams 1979).
2 York (Kenward & Hall 1995).
3 Mummy (Hope 1836).
4 Neuss, Rhineland (Koch 1971).
5 Mummy of Ramses II, Thebes (Alluaud 1908).
6 "in Egyptian mummies" (Attia & Kamel 1965).
7 Mummy of Pum II, Thebes (Cockburn *et al.* 1975).
8 Wilsford, Wilts. (Osborne 1989).
9 Mummy of Han-Em-Ken-Esi (Strong 1981).
10 Changsha, Hunan (Chu & Wang 1975).
11 Runnymede, Surrey (Robinson 1991).
12 Boston, Mass. (Bain 1998).
13 Lea Marston, Warks. (Osborne 1974).
14 Tuna el-Gebel mummified ibises (Boessneck 1988; Levinson & Levinson 1994).
15 York (Hall *et al.* 1980).
16 London (de Moulins 1990).
17 Karanis (Blaisdell 1927).
18 Mandalo, Macedonia (Valamoti & Buckland 1995).
19 Tomb of Tut'ank hamun (Zacher 1934).
20 Herculaneum, Naples (dal Monte 1956).
21 Santorini (Panagiotakopulu *et al.* 1991).
22 Mons Claudianus (Roman) (Panagiotakopulu & van der Veen 1997).
23 Sweet Track, Somerset (Neolithic) (Girling 1984).
24 Southampton (Grove 1995).
25 Deir el Medina, Offering in tomb of Kha (XVIII Dynasty) (Levinson & Levinson 1994).
26 Tomb of Tut'ank hamun (Alfieri 1931).
27 New Kingdom tomb, Thebes (Audouin 1835).
28 York (Roman) (Buckland 1976b).
29 Aachen (Roman) (Koch 1971).
30 Alcester, Warks. (Roman) (Osborne 1971).
31 Workmen's Village, el-Amarna (Panagiotakopulu, this vol. & in prep.).
32 York (Roman) (Hall & Kenward 1990).
33 Offering in tomb (VI Dynasty) (Andres 1931).
34 St Albans, Herts. (Roman) (Bradley 1958).
35 Troy (Panagiotakopulu, this vol.).
36 Meare, Somerset (Neolithic) (Caseldine 1987).
37 Dragonby, Lincs. (Roman) (Buckland 1996).
38 Göttingen (Neolithic) (Büchner & Wolf 1997).
39 Servia, Macedonia (Neolithic) (Hubbard 1979).
39 Tel Arad, Negev (Iron Age) (Hopf & Zachariae 1971).
40 Saqqara, tomb of King Djoser (III Dynasty) (Levinson & Levinson 1994).

Table 3 -3 The earliest records of beetle-pests in the storeroom.

Note :

(Bain, pers. comm. & 1997) has recently found *Cercyon analis* (Payk.), *Creophilus maxillosus* (L.), *Quedius mesomelinus* (Marsh.), *Oryzaephilus surinamensis* (L.), *Bruchus* cf. *pisorum* (L.) and *Sitophilus granarius* (L.) from deposits of AD 1621-1673 at Ferryland, Newfoundland.

4. Early references to insecticides

Introduction

The following chapter is a summary of the information found in Egyptian and Classical texts concerning insect pests and the ways to control them. From the second millennium BC Egyptian Papyrus of Ebers to Byzantine texts, agricultural life can be considered from the insect point of view.

The Papyrus of Ebers

The oldest reference to treatment against insect and other storage pests, known from the Mediterranean region comes from the Papyrus of Ebers from Egypt (Ebell 1937). It was found in 1862 together with what is known as the Edwin Smith papyrus and it was purchased some years later by Ebers. The Ebers papyrus was written during the XVIIIth Dynasty, but there is evidence that it contains parts of at least forty different books, many centuries older than the papyrus itself (Glanville 1942, 180). It is dated to *ca.* 1600 BC, and includes magical recipes for different purposes, some of which have medicinal value. It is currently kept in the University of Leipzig Library.

The translation of many technical terms in such documents is incomplete and their meanings remain unknown. It does, however, provide interesting data on pest control. The papyrus shows that the ancient Egyptians were aware of insect pests in the houses as well as in the storerooms, and took measures against them. It contains a series of remedies for expelling fleas from the house, such as sprinkling natron water (sodium carbonate solution), coating the house thoroughly with **bbt**-plant ground with charcoal, application of natron or of an onion to the holes where the insects live. Against flies, the fat of the **gnw**-bird is recommended, against gnats fresh "*balanites*" (acorn?) oil, while to keep away mice, cat's fat. Gazelle dung ash was used to fumigate barns, and was mixed with water to cover floors and walls; "that is the way to prevent the corn being eaten". A scientific explanation for the above treatments has been proposed (Levinson & Levinson 1985): "the repellence induced by rancid cat or bird fat (applied to sacks) was due to the release of short-chain fatty acids following hydrolysis, while the respective action of dung ash (applied to the walls and floors) depended on the irritant effect of some mineral salts." However, it is more likely that the cat and bird fat recipes are based on sympathetic magic: cat fat against mice and bird fat against insects, for understandable reasons.

Another small section of the papyrus describes fumigating the house and the clothes with aromatic plants, such as dry myrrh, pingon, frankincense, rush-nut, wood of cinnamon, reed from Phoenicia (*Calamus aromaticus*) and liquid styrax (a balsam obtained from the tree *Liquidambar orientalis* (Lucas 1948, 116)) ground fine, mixed together and placed over a fire. Aromatic plants, including spices, were widely used by the Egyptians. Colocynth, coriander, cumin, dill and galingale were used for medicinal purposes together with figs, onions, pomegranates, castor oil, honey, and even certain resins (Harris 1971).

Spices were also known and used at Bronze Age Knossos, Mycenae, and Pylos (Fig. 4.1), as evidenced by several Linear B tablets (Fig. 4.2, 4.3). Linear B and Linear A are scripts that were in use by the Minoans and the Myceneans in Crete and mainland Greece during the Bronze Age. They were thus named by the archaeologist Arthur Evans, so as to differentiate them from hieroglyphics. Each one is a syllabary and they were written with a sharp point on clay tablets. Linear A tablets have been found in Middle Minoan to Late Minoan contexts; it has yet to be deciphered. Linear B tablets have been found in Late Minoan II deposits in Crete and Mycenaean III A-B on the mainland (1500-1100 BC); the script was deciphered by Ventris in 1952 (Chadwick 1958), and clearly represents an archaic form of Greek. Among the spices recorded are coriander [*ko-ri-ja-do-no*] or [*ko-ri-ja-da-na*], red safflower [*ka-na-ko e-ru-ta-ra*], white safflower [*ka-na-ko re-u-ka*], cummin [*ku-mi-no, ku-mi-na*], fennel [*ma-ra-tu-wo*], mint [*mi-ta*], sesame [*sa-sa-ma*], cyperus [*ku-pa-ro*], cardamon [*ka-da-mi-ja*], celery [*se-ri-no*], ?ginger grass [*ko-no, ko-i-no*], and caraway [*ka-mu-nu*] (Ventris & Chadwick 1973, 221-231). Although archaeological evidence suggests the use of some of these as insecticides (Panagiotakopulu *et al.* 1995), the bare list of commodities provided by the texts gives no information with regard to usage or the degree of loss to pests of the various other foodstuff recorded in the tablets.

Homer is similarly devoid of useful comments upon pests directly, but does refer to sulphur as a fumigant. Odysseus asks from Eurycleia, after killing his wife's, Penelope's, suitors on Ithaca:

> "Οἶσε θέειον, γρηῦ, κακῶν ἄκος,
> οἶσε δέ μοι πῦρ,
> ὄφρα θεειώσω μέγαρον·"
>
> (Homer XXII 490-494)

> "Now, old nurse, bring sulphur to cleanse away this pollution; bring fire as well so that then I may purify the house."
>
> (trans. by Shewring 1984).

During the New Kingdom in Egypt, sulphur was similarly applied as dust or as fumigant in the granaries against pests (Levinson & Levinson 1985). Finally the help of gods was requested for the protection of the storerooms. The god Horus gave protection against harmful animals, and Anubis, Khepri, Khnum, Mut, Osiris, and Toth were also worshipped for the same reason (Bodenheimer 1928).

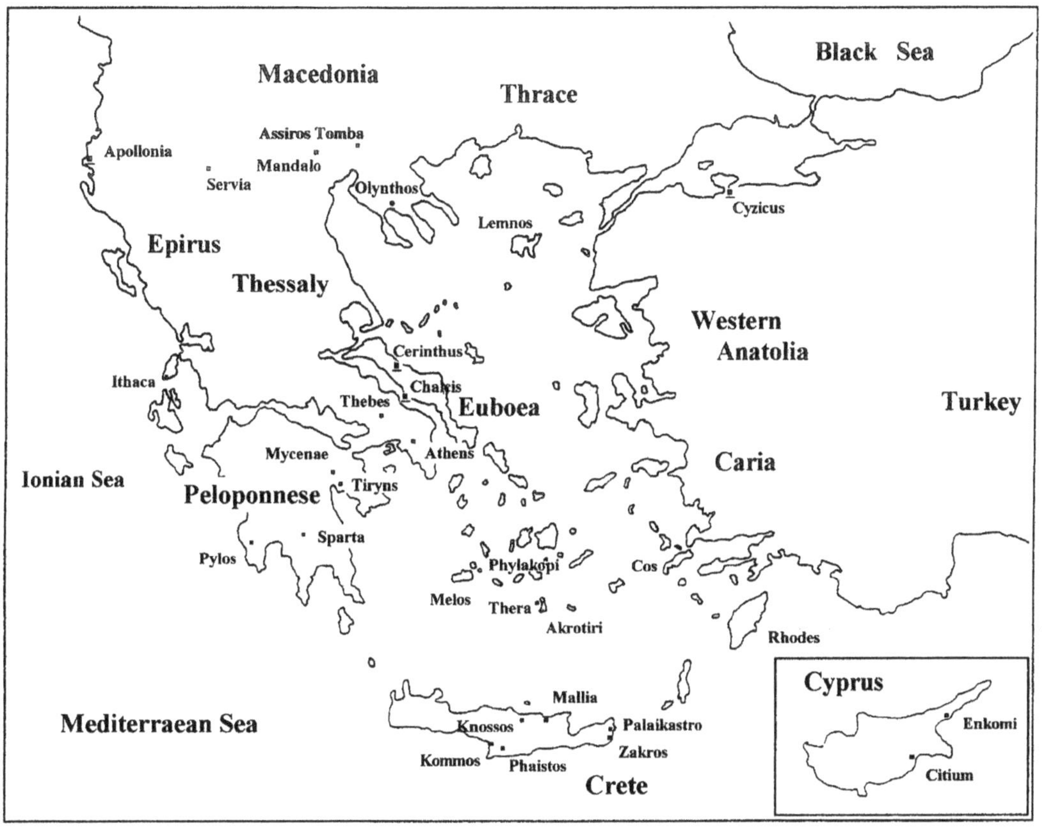

Fig. 4.1 Map of Greece with sites mentioned in the text (■ = site mentioned in Classical sources)

The Classical sources

Several writers during the Classical period made records of agricultural life, leaving us with an invaluable source of information about field cultivation, storage, agricultural implements and everyday activities. Cato, Columella, Pliny the Elder, Varro and Theophrastus all wrote books dedicated to agriculture, and these have been effectively considered by White (1970), amongst others.

Concerning insect pests

Field and storage pests were very well known during Classical antiquity. Although different ways of treating them are quoted in agricultural books, many aspects of their nature remained unsolved. The Classical agricultural writers knew about pest infestation, without being able to distinguish different pests. The name "κίς" is used most of the time for all of the pests, and most probably it originated from a group of monosyllabic names referring to small insect pests: κόρις, φθείρ, σής, ἴπς, θρίψ (Beavis 1988). Theophrastus, writing in the fourth century BC, uses the name κίς only for the pests of wheat and barley, while the pest in beans is called μίδας (CP IV 15.4). In HP VIII 11.2, he names all pests of cereals and pulses σκώληκες. The pest in beans is also named τρώξ (Photius, *Strattis* fr.80, Em 770.43, Hesychius, *Tzetzes ad Hesiod* Op 418) from τρώγω, eat and κοίελος (Hesychius).

The Latin name for κις is *curculio* (Georg I 186, C.GI.L. iii 281.19). The names, φθείρ, louse, and σκώληξ, or vermin, worm, are used even today among Greek peasants (η ψείρα, το σκουλήκι). The clothes moth was named by the Greeks σής, and by the Romans *tinea.* Linnaeus, however, applied the name *Pthirus* to the crab louse, *P. pubis*, and *Tinea* to the genus including the clothes moth.

Descriptions of Pest Infestation

> "Γίνεται δὴ φθειρομένοις σπέρμασι ἴδια ζῶα, καθάπερ ἐλέχθη, πλὴν ἐρεβίνθου· μόνος γὰρ οὗτος οὐ ζωογονεῖ. καὶ σηπομὲνοις μὲν πᾶσι σκώληξ κοπτομένοις δὲ καθ' ἕκαστον ἴδιον. πάντων δὲ μάλιστα διαμένουσιν ἐρέβινθος καὶ ὄροβος, τούτων δ' ἔτι μᾶλλον ὁ θέρμος· ἀλλ ἔοικέ γ' οὗτος ὥσπερ ἀγρίῳ."
>
> (HP VII 6.1)

> "Again, as seeds decay, they engender special creatures, except chickpea, which alone engenders none. As they rot, all produce a grub; but as they get worm-eaten, each produces a special creature. Chick pea and vetch keep best of all, and better still than these lupin; but this, as it were, is like a wild kind."
>
> (trans. by Hort 1916).

The Classical texts reveal, not surprisingly, ignorance about the reproduction of insect pests; they believed in spontaneous generation. Pliny in the first century AD says:

> *"in tritico maxime nascuntur animalia, quoniam spissitate sua concalescit et furfure crasso vestitur."*
>
> (HN XVIII LXXIII,304)

> "It is specially in wheat that grubs breed, because its density makes it get hot and the grain becomes covered with thick bran."
>
> (trans. by Rackham 1950).

Based on the ancient sources and the archaeological record of pests, we can make assumptions about the identity of the insect pests. *Sitophilus granarius* (L.), is still the most common pest in cereals in the Mediterranean. The archaeological record shows that it was the same thousands of years ago. Κίς or *curculio* appears to have been the common name for the weevil in grain in Classical Greek and Roman times. The Latin name *curculio* is used in entomology to describe all weevils, the family Curculionidae. *Oryzaephilus surinamensis* (L.), *Rhyzopertha dominica* (F.), *Stegobium paniceum* (L.), *Lasioderma serricorne* (F.), and *Tribolium castaneum* (Hbst.) are all recorded from early sites around the Mediterranean basin (e.g. Alfieri 1931; Panagiotakopulu & van der Veen 1997), but, as even today in Akrotiri on the island of Santorini, all are described with the same names, η ψείρα, the 'louse', or το πουλί, the 'bird', it is not possible todifferentiate them in the sources. The pest in pulses, Theophrastus' μίδας, most probably referred to species in the family Bruchidae. Bruchids are the main pests in pulses and very common pests in Eastern and Southern Europe. *Bruchus rufipes* Hbst., the same pest that infested pulses in Late Bronze Age Akrotiri, is found in the storerooms of the village today, referred to by the farmers as "το πουλί στη φακή", the bird in the lentils.

Observations on insect food habits

"πονεῖν δ' ἔνια φαίνεται ταῖς ὀσμαῖς καὶ ταῖς εὐωδίαις, εἴπερ ἀληθὲς τὸ ἐπὶ τῶν γυπῶν καὶ τῶν κανθάρων. τοῦτο δὲ δῆλον ὡς δι' ἐναντίωσιν τῆς ἐν αὐτοῖς φύσεως. ὡς δὲ καθ' ἕκαστον ἅμα δεῖ τὴν γε κρᾶσιν τὴν ἑκάστου καὶ τὴν τῆς ὀσμῆς λαμβάνειν δύναμιν."

(Theophr. O 4).

"indeed some animals seem to be annoyed by odours, even good ones, if what is said of vultures and beetles be true; the explanation is that their natural character is antipathetic to odours. To appreciate this in particular cases one should take into consideration the temperament of the animal in question and also its power of smell."

(trans. by Hort 1916).

The above paragraph summarises the principle behind the recipes against insects. In CP (IV 16.2) Theophrastus expands more on insect food habits:

"τὸ δ' ὅλον ἄκοπα τὰ μὲν ξηρότητι (διὸ καὶ ἡ κάχρυς χρονιώτερον), τὰ δὲ ἀμιξίᾳ καὶ καθαρότητι (καθάπερ ὁ χόνδρος), τὰ μὲν χυμῶν φύσει καὶ δυνάμει φυσικῇ (καθάπερ ἐρέβινθος θέρμος σήσαμον), τὰ δὲ τῷ πολὺ τὸ αποστέγον ἔχειν (ὥσπερ ὁ ὦχρος· ἅμα δὲ τοῦτό γε καὶ ξηρόν καὶ θερμόν)."

" In general some seeds escape getting worm-eaten by reason of their dryness (which is why parched barley keeps longer) whereas others do so by staying clear of mixture and remaining pure (like groats), some doing this by the nature of their flavour and their natural power, as chickpea, lupine and sesame, others by their thick insulation against the outside, as bird's peas (but this seed is at the same time also dry and hot)."

(trans. by Einarson & Link 1990).

Pungency seems to be another insecticidal property mentioned by Theophrastus. The use of wine for soaking seeds before sowing is explained because wine has "a certain pungency and pungencies are preservative." (CP III 24.4). According to Theophrastus, there are some qualities in different seeds that reduce the danger of infestation, such as odour, flavour, content of humidity, purity of crop, natural power of the seed, insulation of the seed, pungency (CP IV 16.2, III 24.4).

Climate and geographic position as components in the preservation of crops

"πρός ἅπαντα δὴ ταῦτα μεγάλα διαφέρουσιν αἱ χῶραι οὐκ ἀλόγως· ὁ γὰρ ἀὴρ εὐθύς διάφορος τῷ θερμὸς ἢ ψυχρὸς εἶναι ἢ ὑγρός ἢ ξηρός· οὗτος δ' ἦν ὁ γονεύων· δι' ὃ καὶ ἐν οἷς εἰώθασι γίνεσθαι οὐκ ἀεὶ γίνονται."

(HP VII X.5)

"Now in regard to all these pests the position makes a great difference, as might be expected. For the climate, it need hardly be said, makes a difference according as it is hot or cold, moist or dry; and it was the climate which gave rise to these pests; therefore they are not always found in places in which they ordinarily occur."

(trans. by Hort 1916).

So, the quality of crop, the duration of storage, the preservation state are defined by the climatic conditions and the geography of the area concerned. To the above reasons can be attributed the general belief that,

"ἐν Ἀπολλωνία γοῦν τῇ περὶ τὸν Ιόνιον οὐκ ἐσθίεσθαί φασιν ὅλως κύαμον, δι' ὃ και εἰς θησαυρισμὸν ἀποτίθεσθαι; διαμένει δὲ καὶ περὶ Κυζικὸν ἐπὶ πλείω."

(HP VIII XI.3)

at Apollonia on the Ionian sea and around Cyzicus,

"beans do not get worm-eaten at all"

and that,

"τάχιστα δ' ὁ κύαμος, θερισθεὶς γὰρ εὐθὺς καὶ συντεθεὶς ἐνιαχοῦ, καθάπερ ἐν Θετταλίᾳ."

(CP IV 16.1)

"beans get worm-eaten fastest, for in some countries (as in Thessaly) this happens directly after harvesting and storage."

(trans. by Einarson & Link 1990).

The direction of the wind was also seen as essential. Columella (RR II XX.5) advises a west wind when corn is harvested; Pliny, however, finds it too dry. Columella's comment may has more to do with him being brought up in Spain rather than in Italy, or perhaps Pliny's primary source - much of his data are derivative - provided different information on harvest.

General methods for pest control

The struggle against insects does not start in the storerooms, but actually well before sowing. The choice of the plant, even of the very seeds to be sown, was known to be a critical issue. Surprisingly, the basics of biological control were well known to Classical Greek and Latin writers. The natural power of the seed is the most important quality of a good crop (CP IV 16.2, IV 2.2). "Achillean barley", for example, was believed to be more resistant to bad conditions than other kinds of barley (HP XIII 10.2). The requirements of the farmer include also "the longevity of the seeds" (Pliny HN XIX LVIII 181).

"*Nam quod ad permutationem seminum attinet, quibusdam ex his firmitas maior est, ut coriandro, betae, porro, nasturtio, sinapi, erucae, cunilae et fere acribus; infirmiora autem sunt atriplici, ocimo, cucurbitae, cucumi, et aestiva omnia hibernis magis; minime autem durat gethym. sed ex his quae sunt fortissima nullum ultra quadrimatum utile set, dumtaxat serendo: curilis et ultra tempestiva sunt.*"

(HN XIX LVIII 181)

"In regard to the deterioration of seeds, some keep longer than others, for instance coriander, beet, leek, cress, mustard, rocket, savory and the pungent seeds generally; while the seeds of citrus, basil, gourd and cucumber do not keep so well, and summer seeds in general are not so strong as winter ones. The least lasting is long-onion seed. Of these however which keep best none is of any use even beyond that period."

(trans. by Rackham 1950)

and,

"*Diuturnitatiis causae plures: aut in ipsius grani corio cum est numerosius, ut milio, aut suci pingue-dine, qui pro umore sufficiat tantum, ut sesimae, aut amaritudinae, ut lupino et cicerculis. In tritico maxime nascuntur animalia, quoniam spissitate sua concalescit et furfure crasso vestitur.*"

(HN XVIII LXXIII 304)

"There are several causes that make grain keep: they are found in the husk of the grain when this forms several coats, as with millet, or in the richness of the juice, which may be enough to supply moisture as with gingelly, or in bitter flavour, as with lupine and chickling vetch. It is specially in wheat that grubs breed, because its density makes it get hot and the grain becomes covered with thick bran."

(trans. by Rackham 1950).

Locusts could mean famine and death in the Classical period. Livy (*Epit.* 60) mentions the death of 200,000 people in the area of Carthage and Utica after an attack of locusts in 125-124 BC. An extraordinary reference to the confrontation of plagues of locusts in different parts of the Roman Empire and the world is given by Pliny,

"*Italiam ex Africa maxime coortae infestant, saepe populo Romana ad Sibyllina coacto remedia confugere inopiae metu. in Cyrenaica regione lex etiam est ter anno debellandi eas, primo ova obterendo, dein fetum, postremo adultas, desertoris poena in eum qui cessaverit. et in Lemno insula certa mensura praefinita est quam singuli enecatarum ad magistratus referant. graculos quoque ob id colunt adverso volatu occurentes earum exitio. necare et in Syria militari imperio coguntur. tot orbis partibus vagatur id malum; Parthis et hae in cibo gratae.*"

(HN XI XXXV 105-106)

"Italy is infested by swarms of them, coming principally from Africa, the Roman nation having often been compelled by fear of dearth to resort to remedies prescribed by the Sibylline Books. In the district of Cyrene there is actually a law to make war upon them three times a year, the first time by crushing the eggs, then the grubs and last the fully grown insects, with the penalty of a deserter for the man who shirks. Also in the Island of Lemnos there is a rule prescribing a definite quantity of locusts killed that each man has to bring in to the magistrates. Also they keep jays for this purpose, which meet them by flying in the opposite direction to their destruction. In Syria as well people are commanded by military order to kill them. In so many parts of the world is this plague abroad: but with the Parthians even the locust is an acceptable article of diet."

(trans. by Rackham 1940).

Treatments before sowing

Treating the seed before sowing was believed to make it less liable to disease. Some recommend wine (HN XVIII XLV, CP III 24.4) or a concoction of houseleek dip (Democritus in HN XIX XVIII 159). Virgil proposes a soak of nitre and black oil lees for beans (Georg. I 193-4), while others recommend a mixture of urine and water, three days before sowing; hoeing three times is the best method to obtain the best seed for sowing. Cypress leaves when mixed with seeds diminish the threat of "maggots" (HN XVIII XLV 158).

Pliny also recommends sowing by the new moon and burying a toad in the middle of the field (HN XVIII XLV 158). The moon was thought to influence all agricultural activities, while the toad was supposed to scare the pests away. He further comments that it should not be left for long in the ground or it would frighten the soil also! When the seeds are sprinkled with ashes of a weasel or a cat, or with water in which mice have been boiled, mice are driven away from the field. However an ox-gall dip is preferable, since the others leave a bad odour even in bread (HN XVIII XLV 160). These recipes are obviously based on prejudice. However the use of ashes in pest control, even if there is a magical ingredient, is known to be efficacious.

In the field and the garden

Field infestation can ruin an entire crop in a short period of time. Birds, mice, insects, worms, and plant diseases all act against the welfare of the farmer. In Antiquity, a variety of field treatments were employed to ensure the quality of the crop yield and to keep away dangerous visitors. For example, "pitch, oil and grease are particularly effective for young trees" (HN XVII XXXVII 234); and these techniques are still used by modern gardeners and fruiterers (e.g. Hay 1972). Caterpillars in cabbages were killed by sprinkling wormwood (*Artemisia* sp.) extract or houseleek on them (HN XIX LVIII 179). Fumigation of damp gardens by burning *galbanum*, a fragrant gum resin (Lucas 1948, 115), was also re-commended against gnats (HN XIX LVII 180). In addition, magical rituals are mentioned for protection against insects and caterpillars. These include use of a frightening skull of an equid or a river crab, or touching plants with blood-red cornel.

Threshing floors

"*Aream sic facito. Locum ubi facies confodito. Postea amurca conspargito bene sinitoque condibat. Postea conminuito glebas bene. Deinde coaequato spargito sinitoque arescat. Si ita feceris, neque formicae nocebunt neque herbae nascentur.*"

(DA XCI)

The construction of a threshing floor was usually one of the last activities to break the ground. It was recommended that it be soaked thoroughly with *amurca*, a mixture of oil lees and soda, let it be absorbed and afterwards levelled thoroughly (DA *idem*). Other sources recommend that the floor be merely levelled and covered with a weak solution of water and cowdung (HN XVIII, LXXI), and, because:

"*saepae exiguus mus sub terris
posuitque domos atque horrea fecit, aut
oculis capti fodere cubilia talpae,
inventusque cavis bufo et quae plurima
terrae monstra ferunt, populatque
ingentem farris acervum curculio atque
inopi metuens formica senectae.*"

(Georg. I 178)

"often under the ground the tiny mouse sets up a home and builds his storehouses, or sightless moles dig out chambers; in holes may be found the

toad, and all the countless pests born of
the earth: or the weevil savages a huge
heap of grain, or the ant anxious for a
destitute old age."

(trans. by Lewis 1940).

Virgil proposes dressing the threshing floor with binding clay.

Storage

"ὧδέ κεν ἁδροσύνῃ στάχυες νεύοιεν ἔραζε,
ἐς τέλος αὐτός ὄπισθεν 'Ολύμπιος ἐσθλόν ὀπάζοι,
ἐκ δ' ἀγγέων ἐλάσειας ἀράχνια· καί σε ἔολπα
γηθήσειν βιότου αἱρεύμενον ἔνδον ἐόντος.
εὐοχθέων δ' ἵξεαι πολιὸν ἔαρ, οὐδέ πρὸς ἄλλους
αὐγάσεαι· σέο δ' ἄλλος ἀνήρ κεχρημένος ἔσται.."

Hesiod, *Works and Days* (473-8)

"If this way your corn ears will bow to the ground
with fullness if the Olympian himself gives a good result
at the last, and you will sweep the cobwebs from your bins
and you will be glad, I ween, as you take of your garnered substance. And so you will have plenty
till you come to grey springtime, and will not look
usefully to others, but another shall be in need of your help."

(trans. by White 1929)

In communities continuously threatened by famine and malnutrition, effective storage techniques were a priority for both foodstuffs and for seed for the following year. Good conditions of preservation of stored crops could secure the balance of life for the settlement. The depth of knowledge of conditions during storage is attested by several Classical agricultural writers. The first principle for long term preservation of a crop was that the seed to be kept should be gathered dry, so that there was less moisture in it and mould was less likely to attack the food. Also, harvesting the grain at the right time made a great difference in terms of effective storage:

"*nobis referre plurimum tempestivitas condendi videbitur, nam si parum tostum atque robustum collectum sit aut calidum conditum, vitia innasci necesse est.*"

(HN XVIII LXXIII 303)

"To us storing the corn at the proper time will seem most important as if it is got in when insufficiently ripened and firm, or stored while hot, pests are certain to breed in it."

(trans. by Rackham 1950).

An issue that seems to leave the Classical agricultural writers in disagreement was whether the grain in the storerooms should be aired or not.

"*multi ventilare quoque vetant: curculionem enim non descendere infra quattuor digitos, nec amplius periclitari.*"

(HN XVIII 302)

"Many moreover forbid turning the grain to air it, as the weevil does not penetrate more than four inches down and beyond that, the grain is in no danger."

(trans. by Rackham 1950).

Columella seems to believe the same as he talks about removing the upper infested portion of the grain and using that underneath.

"*Quod genus exitii cum indicit, multi opinantur arceri posse, si exesae fruges in horreo ventilentur et quasi refrigerentur. Id autem falsissimum est; neque enim hoc facto expelluntur animalia, sed immiscentur totis acervis. Qui si maneant immoti, summis tantum partibus infestentur, quoniam infra mensuram palmi non nascitur curculio; longeque praestat id solum, quod iam vitatum est, quam totum periculo subicere. Nam cum exiget usus, facile est eo sublato, quod vitiatum erit, integro inferiore uti.*"

(RR I VI.16-7).

"Many think that when this kind of pest appears it can be checked if the damaged grain is winnowed in the bin and cooled off, as it were. But this is a most mistaken notion; for the insects are not driven off by so doing, but are mixed through the whole mass. If left undisturbed, only the upper surface would be attacked, as the weevil breeds no more than a palm's breadth below; and it is, far better to endanger only the part already infested, than to subject the whole amount to risk. For it is easy, when occasion demands it to remove the damaged portion and use the sound grain underneath."

(trans. by Ash 1941).

Theophrastus also argues that winnowing is wrong:

"φαίνεται δὲ τοῦτό γε καὶ ἄλλως ἀληθές· ὡς οὐκ εὐθὺς ἀλοαθέντα τὰ σπέρματα βελτίω τῶν χρονισθέντων,..."

(CP IV 6.5)

"This point about imperfection at least appears in any case true: thus seed is not better right after winnowing than after being kept sometime..."

(trans. by Einarson & Link 1990).

Varro (RRm I LXIII), on the other hand, believed that grain, which the weevil had begun to infest, should be brought out for protection. Apparently, the same methods were used for storing cereals and pulses. Most of the time, they are both referred to as "grain", and were stored in storerooms in earthenware jars or in underground silos.

Underground storage

Storing in caves or rock cavities is a very well known technique (Sigaut 1988) that has its origins in early prehistory. The most widespread practice, however, appears to have been storage in pits which were afterwards sealed. There are many references to airtight underground storage in Classical texts.

"*Quidam granaria habent sub terris speluncas, quas vocant sirus, ut in Cappadocia ac Thracia: alii, ut in agro Carthageniensi et Oscensi in Hispania citeriore, puteos.*"

(RRm I LVII 2)

"Some use underground cavities as granaries the so called *sirus* as occur in Cappadocia and Thrace: and still others use wells as in the Carthaginian and Oscenian districts in Hither Spain."

(trans. by Hooper 1935).

Pliny pays testimony to the aforementioned, characterising it as the most appropriate method (*utilissime tamen servantur in scrobibus*). According to both him and Varro, wheat keeps fifty whole years when preserved in this way, while millet may last for one hundred years (HN XVIII LXXIII 307-8, RRm I LVII). Beans, stored in a cavern in Ambracia in Northwest Greece, are alleged to have lasted for a period of 220 years (Pliny HN XVIII LXXIII 307-308).

Varro mentions:

"*Horum solum paleis substernunt et curant ne umor aut aer tangere possit, nisi cuuum promitur ad usum; quo enim spiritus non pervenit, ibi non oritur curculio.*"

(RRm I LVII 2)

"They cover the bottom of these (the wells used for storage) with straw, the weevil does not breed where air cannot reach."

(trans. by Hooper 1935).

Elsewhere, Varro, describing the preservation of pomegranates, gives us another method of underground storage:

"*eaque si coieceris in terram et obleris circum rarum, ne extrinsecus spiritus adflet...*"

(RRm I LIX)

"if you put them on the branch, put them in a pot with no bottom, bury them in the ground and seal the end of the branches so that no outside air will reach them."

(trans. by Hooper 1935).

Building storerooms

The need for security and good preservation of foodstuffs dictated the construction of rooms or buildings dedicated to crop storage. In the Classical period, the necessity of provisioning both cities and large, standing armies, led to granaries being built on a massive scale (Rickman 1976), and pest infestation problems were exacerbated as a result of large scale transportation of commodities (Buckland 1981).

As noted in Classical sources, the storerooms had to meet a number of criteria:

a. if no air is allowed to penetrate, it is certain that no pests will breed in the grain

(HN XVIII LXXIII 306).

b. decomposition is produced by heat and involves liquefaction

(CP IV 16.3).

Describing the architecture of the farmhouse, Vitruvius refers to the fact that rooms for grain should be set in an elevated position with a north or north eastern exposure. In this way the grain, because of the wind, keeps cool, and therefore no weevils spoil it (A VI 4). In Hither Spain and in Apulia, granaries were built in the field above ground, so constructed that the wind could cool them not only from the sides, through windows, but also beneath from the ground (RR I LVII 3). Columella describes storage in lofts, ventilated through small windows opening to the north. In fruithouses, shutters are recommended (RR I LIX 2). Pliny describes the typical Egyptian granary (HN XVIII LXXIII 301) as made of thick bricks, with no windows, and filled from above.

The interior of the storeroom is equally important. Varro (RRm LVII 1) recommends that the walls and floors are coated with marble cement or at least with grain chaff and *amurca,* as this both keeps out mice and worms and makes the grain more solid and firm. Some farmers sprinkled the wheat too with *amurca,* using a quadrantal to about a thousand modii. *Amurca* (from the Greek word ἀμόργη) is composed of the lees from the preparation of olive oil.

Cato stresses the importance of treating the interior of the storeroom:

"*Frumento ne noceat curculio neu mures tangant. Lutum de amurca facito, palearum paulum addito, sinito macerescant bene et subigito bene; eo granarium totum oblinito crasso luto. Postea conspargito amurca omne quod utaveris. Ubi aruerit, eo frumentum refrigeratum conditio; curculio non nocebit.*"

(DA XCII)

"To keep weevils and mice from injuring grain make a slime of *amurca* with a little chaff added, leaving it quite thin and work thoroughly. Cover the whole granary with the thick slime and then add a coat of *amurca* over the whole. After it has dried, store cooled grain there and the weevils will not injure it."

(trans. by Hooper 1935).

Another way of making the storeroom less liable to infestation was soaking the floor with *amurca* lees and overlaying it, after being dried, with tiles of sand lime and oil lees. The same was recommended to be applied to the joints of the walls. As for the walls, they should be coated with a plastering of clay and oil lees with dried wild olive leaves (Columella RR I VI.14). Theophrastus believed that dusty grain kept in plastered rooms perished more rapidly than that which was kept in unplastered rooms (HP VIII 11.1). Cato, however, thought that careful plastering was necessary, not only for the storeroom, but also for the whole house:

"*Terram quam maxime cretosam vel rubricosam, eo amurcam infundito, paleas indito. Sinito quadridum fracescat. Ubi concideris, delutato. Ita neque aspergo nocebit, neque hebra nascetur, neque lutamenta scindent se.*"

(DA CXXVIII)

"Take very chalky or red earth, pour *amurca* over it and add chopped straw; let it soften for four days and when it has softened thoroughly, work up with a spade and when you have worked it up, plaster. With this treatment, the moisture will not injure the walls, nor the mice burrow them, nor weeds grow, nor the plaster crack."

(trans. by Hooper 1935).

Insecticides and insect repellents

Storage in Antiquity involved also a number of substances, such as ash, minerals, clays, different kinds of soils, sand, salts, oils, animal and plant substances, which were thought to have pesticidal properties. Thanks to the Classical and Byzantine agricultural writers, an extensive record of insecticides and insect repellents has survived.

Mixing earth with the stored grain is recommended by Theophrastus (HP VIII 11.7) who speaks specifically about an earth, γῆ, from Olynthos and Kerinthos. Pliny also proposes the same kind of earth as well as Carian and Chalcic chalk (HN XVIII LXXIII 304). Varro notes:

> "*Item alius aliut adfriat aut aspargit, ut Chalcidicam aut Caricam cretam aut absinthium, item huius generis alia.*"
>
> (RR LVII)

> "Different farmers use different powders or sprays such as Chalcidian or Carian chalk..."
>
> (trans. by Hooper 1935).

Also Plutarch mentions the use of clay mixed into the grain (*Quast. Conv.* V.5.1). Apparently the principle behind the use of clay was its taste;

> "καὶ ἔνιαι γλυκεῖαι δοκοῦσιν εἶναι (καθάπερ καὶ ἡ ἄργιλος);"
>
> (CP VI 3.2)

> "some earths are considered sweet (as clay)"
>
> (trans. by Einarson & Link 1990).

Several minerals do have insecticidal properties. When finely crushed, the sharp edges of crystals may be damaging to insect cuticular joints, leading to dehydration. Gypsum, in particular, has been used in this context. Several authors note a preference for storage in earthenware (DA XCIX, Columella RR XLIV) or clay vessels (Varro RR I LVIII). These are recommended to be made airtight, and sealed (DA CI) either with plaster, gypsum (Columella RR II X.16), or with pitch (DA CVII). Preservation of fruits in sand is also noted:

> "*mala punica demissis suis surculis in dolio harenae...*"
>
> (RR I LIX)

> "pomegranates are preserved by burying their stems in a jar of sand..."
>
> (trans. by Hooper 1935).

Salt was also used as a preservative especially for meat (DA CLXII). Many insects have limited ability to deal with salt, and jars previously used for salt-fish were preferred for storage of legumes ((Columella RR II X.16, Palladius VIII 3.2). Mixing ash in the stored crop or covering it with ash was believed to be lethal to insects (HN XVIII LXXIII 308, Columella RR II X.16, Varro RR I LVII), and would have acted in the same way as mineral matter. Theophrastus ascribes its properties to the flavour that ashes acquire (CP VI 3.5).

Pitch and bitumen were used as preservatives, mainly as a final coating over the treated stored product; Pliny, for example, refers to pitch (HN XVIII LXXIII 308). Cheese, according to Columella (RR XII XLIII), was coated with pitch so as to keep longer and better, and grapes were also treated with it (Columella RR XII XLVI). As noted above, pitch was used as an ingredient for the treatment of young trees to avoid pest infestation and disease (HN XVII XXXVIII 234), and bitumen was applied to vines for the same reasons (DA XCV, Columella RR XII XIV).

Dung was put on leeks for protection in the field (HN XIX LVII), and sometimes, in the form of a solution, it was used to cover threshing floors (Georg. I 178). Nitre, *nitrum*, is also mentioned (Georg. I 193-194), mixed with *amurca*, for treating seeds before sowing. Apparently, *amurca*, was one of the most effective substances. Its use in several activities has already been mentioned. For preserving, *amurca* was used for the storage of dried figs (DA XCIX), "myrtle or any other twigs with berries" (DA CI), as well as for steeping new oil jars :

> "*Amurca inpleto dies VII, facito ut amurcam cotidie suppleas. Postea amurcam eximito et arfacito.*"
>
> (DA LXIX)

> "fill them with *amurca*, maintaining a constant level, for seven days; then pour off the *amurca* and let the jars dry..."
>
> (trans. by Hooper 1935).

This may, however, be largely for sealing the somewhat porous, unglazed containers. *Amurca* was also used in the storage of clothes in order to avoid moths:

> "*Amurcam decoquito ad dimidiuum, ea unguito fundum arcae et extrinsecus et pedes et angulos. Ubi ea adaruerit, vestimenta condito. Si ita feceris, tiniae non nocebunt.*"
>
> (DA XCVIII)

"fill them with *amurca*, maintaining a constant level, for seven days; then pour off the *amurca* and let the jars dry..."

(trans. by Hooper 1935).

This may, however, be largely for sealing the somewhat porous, unglazed containers. *Amurca* was also used in the storage of clothes in order to avoid moths:

> "*Amurcam decoquito ad dimidiuum, ea unguito fundum arcae et extrinsecus et pedes et angulos. Ubi ea adaruerit, vestimenta condito. Si ita feceris, tiniae non nocebunt.*"
>
> (DA XCVIII)

> "Boil *amurca* down to one half of its volume and rub it over the bottom, the outside, the feet and the corners of the chest. After it is dry store the clothing and the moths will not attack it."
>
> (trans. by Hooper 1935).

Dry chaff could be placed at the bottom of the vessels, and be mixed with the stored product, or used to cover it (Columella RR XII XLIV.1, Varro RRm LIX). Placing apples on wool (RRm LIX) was an alternative method for accomplishing good preservation. The most popular insect repellent in ancient sources seems to have been olive oil. Theophrastus writes about it:

> "καὶ ὡς τὰ ἔντομα πάντα ὑπὸ τοῦ ἐλαίου βαρύνεται, φεύγει γὰρ αὐτὴν τὴν ὀσμήν·"
>
> (CP VI 5.4)

> "and just as all insects find oil oppressive for they avoid the smell of it."
>
> (trans. by Einarson & Link 1990)

and,

> "ἀλλὰ τούτου μὲν τὴν αἰτίαν ἐν τῇ δριμύτητι ληπτέον, ὥσπερ καὶ τῆς ὀριγάνου καὶ τῶν τοιούτων; ἅπαντα γὰρ φεύγει."
>
> (CP VI 5.4)

> "But the cause in the case of olive oil is to be found in its pungency just as with marjoram and the like, for insects avoid them all."
>
> (trans. by Einarson & Link (1990).

Aelian (IV.18) believed that insects died if oil was poured on them. Cato (DA CXLII), Sextus Empiricus (Pyrrh I.55), and Hippiatrica Cant. (LXX 1-13), Hippiatrica Berol. (LXXXV 4-7), talk about the antipathy of insects to oil and include it as a basic ingredient in recipes against insects. Storage of lentils included rubbing them with oil and then drying them (DA XCV). Olive jars were used for storage of cereals and pulses (Varro RR LVII, Columella RR II X.16), probably because their porous surface was sealed with the remains of oil, leading to suffocation of any pests by carbon dioxide as the seeds respired.

Vinegar was used as a preservative for lentils and meat (DA CLXII), while a soaking of must (*musto*) was the standard way of preserving pears, grapes and cheese (Columella RR LIX LVIII). Perfumes and incenses, especially the perfume of roses, were believed to be lethal too.

Aristotle writes that :

> "ὡσαύτως δὲ καὶ τοὺς κανθάρους ὑπὸ τῆς τῶν ρόδων ὀσμῆς."
>
> DMA 845b.

> "beetles also (die) from the smell of roses,"

and Aelian also believed that a beetle dies if you put roses over it (IV.18; "κάνθαρον δὲ ἀπολεῖς, εἰ ἐπιβάλοις τῶν ρόδων αὐτῷ."), and "μύρῳ κάνθαρος ἀποθνήσκει" "beetles die of perfume" (VI.46).

Quotations on perfume are found in Plutarch's *Moralia* (1058A, 1069A, 710E) as well as in *Pyrrh* (I.55), while Dio Chrysostom reveals the insecticidal qualities of honey (*Orat.* 32.98).

Drying infested grain in the sun and mixing with strong smelling plant substances was a standard technique (Columella RR II X.16). *Silphium*, coriander (Palladius I.19, Diosorides III 71) and marjoram (HN XIX LVIII 178) were used not only as condiments and perfumes but also as insecticides. According to Sabinus Tiro, protection against ants was obtained by stopping the mouths of ant-holes with sea-slime or ash (HN XIX LVIII 178), application of the heliotrope plant, or water in which an unbaked brick had been soaked.

Varro seems to have believed that if bowls of water were placed around infested grain exposed in the sun, the weevils would fall in them and drown (in DA LXIII). The mysterious nature of the weevil and the difficulty in understanding and confronting the damages in different crops made the Romans adopt a god of the storehouse, *Deus Conditor* (Hooper 1935).

Storage of legumes

In Classical texts there are several passages on the storage of leguminous seeds.

> "*Sed et lacibus distiguntur granaria, ut separatim quaque legumina ponantur.*"
>
> (Columella RR I VI 13).

Every different kind of legume "is stored by itself" in granaries that are "divided into bins"

Referring to "*faba et legumina*", Varro (RRm LVII) confirms that they

> "*in oleariis vasis oblita cinere perdiu incolumia servantur.*"
>
> "are kept fresh in olive jars sealed with ashes"
>
> (trans. by Hooper 1935).

According to Columella, lentils would keep for a long time if they were stored mixed with ashes. Another treatment was to sink them into water and separate the infested grains, which float, from the sound ones, dry them in the sun, treat with *silphium* and vinegar, dry again and store either in a big bin or for smaller quantities in olive or salt-fish jars. Sealing the jars with gypsum immediately after they are filled is also recommended.

> "*Ea ne curculionibus absumatur-nam etiam dum est in siliqua exestur-curandum erit, ut cum extrita sit, in aquam demitatur et ab inani, quae protinus inatat, separetur solida; tum in sole siccetur et radice silphi trita cum aceto adspargatur defriceturque atque ita rursus in sole siccata et mox refrigerata recondatur, si maior est modus, in horreo, si minor, in vasis oleariis salsamentariisque; quae repleta cum confestim gypsata sunt, quandoque in usus prompserimus, integram lentim reperiemus. Potest temen etiam citra istam medicationem cineri mixta commode servari.*"
>
> (RR II X 16)
>
> "To keep it from being destroyed by the weevils-for they eat it even when it is in the pod-care must be taken that, as soon as it is threshed out, it be sunk in water, and that the sound grains be separated from the empty, which come at once to the surface; then that it be dried in the sun, sprinkled and rubbed with the bruised root of silphium mixed with vinegar, and again dried in the sun; and presently after cooling, that it be stored away-in the bin if the amount is rather large, or in olive jars and salt-fish jars if there is not much of it. If they are sealed with gypsum immediately upon being filled, we shall find the lentil sound whenever we take it out for use. Still, it can be kept satisfactorily without such treatment if mixed with ashes".
>
> (trans. by Ash 1941).

Cato proposes:

> "*Laserpicium aceto diluito, permisceto lentim aceto laserpiciato et ponito in sole. Postea lentim oleo perfricato, sinito arescat. Ita integra servabitur recte.*"
>
> (DA CXVI)
>
> "Infuse *asafetida* (= silphium) in vinegar, soak the lentils in the infusion of vinegar and asafetida, and expose to the sun; then rub the lentils with oil, allow them to dry, and they will keep quite sound."
>
> (trans. by Hooper 1935).

Fig. 4. 2. Coriander in Linear B (after Ventris 1988).

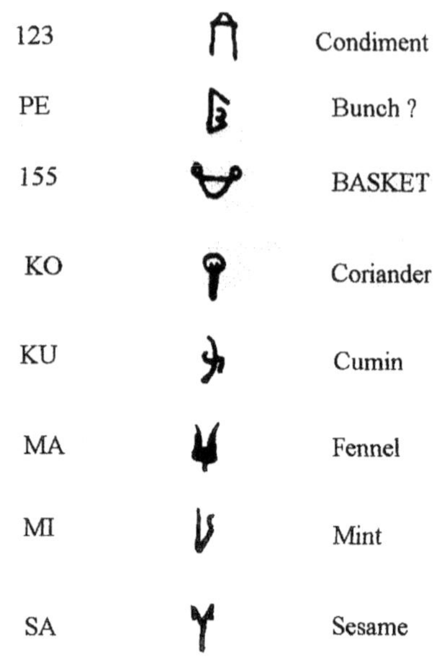

Fig. 4. 3. Spices on Linear B tablets from Mycenae (after Ventris & Chadwick 1973).

Palladius mentions the use of coriander and fleabane (*Erigonum* sp.) foliage for lentils as insecticides (I.19).

Conclusion

It is evident that, from the beginnings of agriculture, production and storage were troubled by insect pests, and several measures were taken to prevent losses of crops. Some of the treatments were merely ceremonies so as to gain Nature's favour, but also a long tradition of pest control existed and was passed on through such documents as the Papyrus of Ebers and the Classical texts. The choice of seed, the climatic and geographic conditions of the area, the treatment of the seeds, before sowing, in the field, and in the storerooms, as well as a deep knowledge about storage architecture, and storage techniques were crucial elements for achieving a good crop, and may have often provided the difference between survival and starvation.

5. Crop Storage

Introduction

The principles behind crop storage are complex and variable (Sigaut 1988). Choice of place, construction methods of storerooms, nature of recipients, choice of commodities stored, and remedies and mixtures applied against pests are some of the critical points needed to be considered before storage. Different ways of storage are practiced around the world (summarized in Sigaut 1978), and the methods of storage relate to the commodities stored, the climate, the environment and the geography of a place. There is also a certain range of operations which need to take place before and during the storage period that are crucial for success in obtaining a good degree of preservation. Whilst an important element in many mixed and pastoral economics, the storage of hay or other fodder crops is not considered directly in this discussion.

Process and choices preceding the storerooms

Drying

Drying of grain and other products is a common method of reducing moisture and controlling insect infestation. It is practiced before storage, and is related to the way the product is stored. The necessity of drying is accentuated by the humidity of climate, the time of harvest and the maturity of the grain (Sigaut 1988, 8). Many insect and fungal infestations are a reflection of local dampness of the stored product, providing nuclei for infestation. In the ethnographic literature, it is apparent that drying usually took place on the threshing floors, and it is unlikely that discrete structures for such processes would be recognised in the archaeological record, although drying on the harvested plant may have utilised frames.

Ways of storing the grain

After drying, grain may be stored in bulk with chaff, in bulk without chaff, in the ears, in the ears with the culms attached, or even in sheaves (Sigaut 1988). Different ways of grain storage have different advantages and disadvantages. For example, grain stored in the sheaves is protected against heat and insect damage, but takes up a larger volume than threshed grain (Sigaut 1988, 7).

Storage methods

Ventilated storerooms

Two principal approaches to long term preservation of commodities are:

i) the development of storage with ventilation,

and

ii) the development of hermetic storage.

The principle of storage with ventilation is the use of the external air to stop overheating and mould growth. This may be carried out either with aeration or with ventilation. Ventilated elevated storerooms have been typical constructions in the Mediterranean at least since the Hellenistic period (Rickman 1971). They are made on a platform above the ground so as to be difficult for rodents, insects and other pests to invade, and also to reduce problems with damp. Wooden or stone discs may also be put around each post to prevent rodents from climbing into the granary (Sigaut 1988).

A typical example of a ventilated storeroom is a "bakhari" (Fig. 5.1), the outdoor granary still found in Bihar, India. Bakharis are elevated from the ground to avoid damp. Their walls are made of Ikri, *Saccharum fuscum*, and plastered with mud. The grain is usually stored surrounded with "bhusa" (broken wheat straw, chaff). The bhusa is well pressed, so as penetration by insects is made as difficult as possible (Bainbridge Fletcher & Ghosh 1921).

The typical Roman military granary form, with huge ventilated bins, has a long history. At Harappa, in the Indus Valley, at around 1500 BC, collective storage was practised, and two great buildings have been identified as granaries. Each of these buildings was divided transversely into six halls, each about 15.8 m by 5 m. Within each granary, the

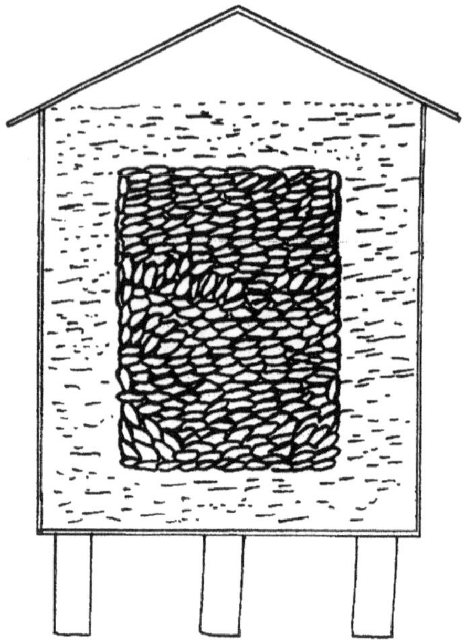

Fig. 5. 1 : Section through a bakhari, showing corn stored surrounded with chaff, Bihar, India (from Bainbridge Fletcher 1921, after Sigaut 1979)

floors were carried upon dwarf walls that were constructed at the same distance from the division walls as from each other, transversely across the building. Ventilators were set in the long sides of the building. Later alterations restricted the size of the ventilators, and additional air holes were constructed so as to ensure the flow of the air (Rickman 1971).

Airtight storage

Pit storage (Table 5.1)

An alternative approach to ventilation is to exclude oxygen in air-tight underground pits, thereby preventing spoilage of grain, since the build up of carbon dioxide, caused by the seeds' respiration, kills any infesting insect pests. The lack of oxygen, however may suffocate people, who enter the pits, if they have not been left open for a while. Different shapes of under-round pits or silos are reported from all over the world: conical, cylindrical, vertical or horizontal, large or small, with varied openings (Fig. 5.2). Sometimes they served only household needs, while at other times they are communal. Pit storage was still practised on a large scale in Sicily in the late medieval period (Bresc 1979, 113-22), and in the seventeenth century on Malta by the Knights of St. John (Reynolds 1974, 119). Today it is still used among the Berbers of the Western Sahara and Northwest Sudan (Nicolaisen 1963, 207). In Africa, pit storage is one of the most common techniques used (Miracle 1966, 53). It was practised in Argentina for large scale grain stores for export (Hall & Hyde 1954, 2), and was until recently practised in Cyprus and the Eastern Mediterranean (Currid & Navon 1989). Nándor's ethnographic work from Hungary (1966) provides a useful review of different types of underground silos. Reynolds (1979) has experimented with pit storage in England, and Currid and Navon (1989) have performed similar experiments in the more arid climate of Palestine.

Europe	
France	Quercy, Vivarais, Roussillon, Gascogne (Gers).
Spain	Catalonia, Tierra de los Barros (Estramadura); villages around Barcelona, Urgel, Tarragona, Burjasot (near Valence), Altafulia, Villafranca, Rota (near Cadiz), Alcala de Gguadayara (near Seville), Monta del Cuervo, Talavera, Zamora, etc.
Italy	Tuscany, Pouilles, Basilicate, Calabria, Sicily, villages around Livourno, Pisa, Arezzo, Florence, Naples, Foggia, San Severo, Vasto, Ortona, Chieti, Termini, Agrigentum, Catane, Brindisi, Tarento, Bari, Otranto, etc. (probably all the villages of South Italy).
Malta	Valetta
Croatia	Dubrovnik
Hungary	throughout
Other countries:	Poland, Czech Republic, Slovakia, Rumania, Bulgaria, Ukraine and Russian Crimea.
Africa	
Libya	The Maghreb
Egypt	Fayum; mentioned in 1947 in oasis and in two small villages near the Nile Delta.
Sudan	
Somalia	
Ethiopia	
Zimbabwe	
Madagascar	Betsileo
Eastern Asia	
Turkey	Aksaray
Cyprus	
Israel, Palestine	
Georgia	
Armenia	
Azerbaidjan	
Iran	
Turkmenia	
Oriental Asia	
China	Sin Kiang
India	Bihar; villages of Ghaziabad, Hapur, Secunderabad, Maharashtra: Sholapur, Poona, Satara, Ahmednagar; Manipur: Tangkhul
Laos	Plateau of Phouens
America	
US	Native Peoples: e. g. Mandan, Hidatsa, Arikara, Pawnee, Omaha; Pueblo (Navaho), Yuma.
Mexico	North-west? Yucatan

Table 5-1 : Geographic distribution of underground pit storage, according to Sigaut (1978; 21, 22).

In different parts of the world, the practice of pit storage is still in use. In Africa, where subterranean storage has been in use for large scale storage at least since the time of the Carthaginian Empire (Gsell 1929, IV: 16-17, Camps 1961, 113-115), the tradition is still continued in the central Sahara. In North Africa, food is stored in individual silos ("matmora"), which are made in dry and solid ground. They are covered inside with clay and fumigated to exterminate any pests. In the Canary Islands, underground silos were constructed in volcanic rocks (Verneau 1891). In Madagascar and in Mali, the practice is still going on for the storage of food commodities (Mauny 1979).

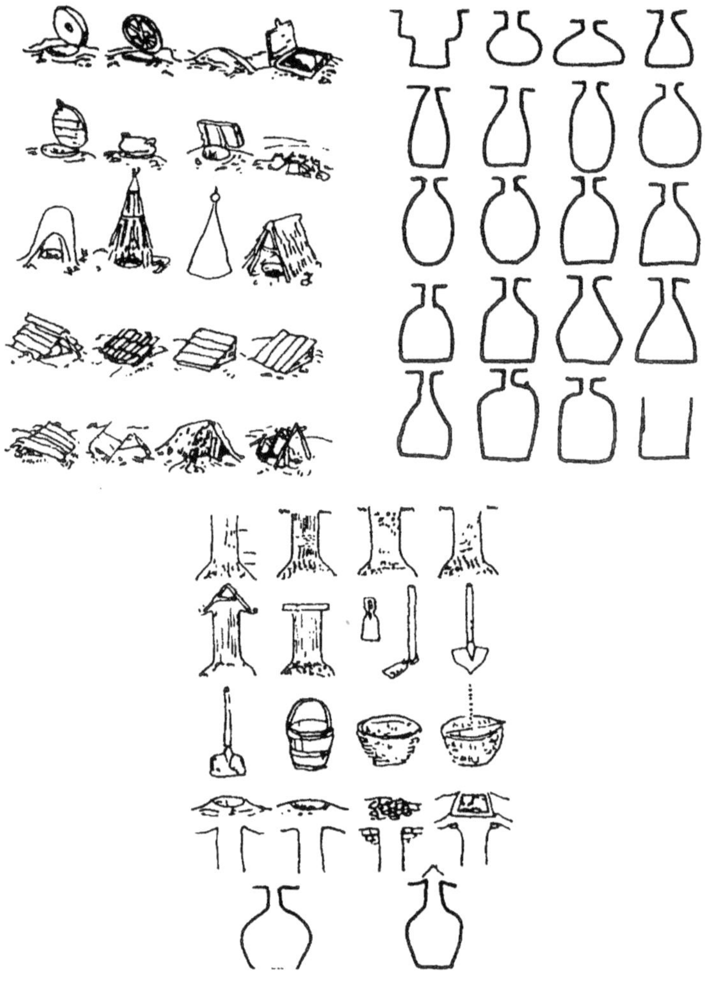

Fig. 5. 2 : Pit closures, pit forms and extraction methods for underground storage of grain (after Nandor (1966), from Sigaut (1988)).

The main defect of this system is that the grain absorbs moisture and ferments to some extent; it is only effective either in semi-arid areas or those with free-draining subsoils. The material closest to the walls becomes black and unfit for human consumption, while the grain may get a strong smell of fermentation. For this reason, grain stored in pits sells cheaper than grain stored in storerooms (Dendy & Elkington 1920), although large scale storage in concrete silos was the recommended method prior to the development of modern insecticides (Munro 1966).

Storage in receptacles

Storage in receptacles is a method of air tight storage, which may be combined with a ventilated building to keep temperatures down. A variety of vessels are used: earthenware jars, wicker bins and baskets, mud bins, bags, hollow tree trunks, etc. The principles are the same as in the case of hermetic storerooms; mud, cowdung or other suitable substances are used to cover the walls and the mouths of the receptacles, so as to ensure their airtightness. Straw or layers of leaf were also used for the same purposes. Ceramic vessels become less porous when covered with resin, pitch, lime or other thick, glue-like mixtures. Ethnographic evidence from the Aegean and Cyprus provides examples of sealing methods applied in storage vessels. In Cyprus, the potters from Phini, who produced large pitharia, used a mixture of hot pitch and turpentine to seal the pot walls (London *et al.* 1990). In Crete, the potters of Thrapsano achieved a similar effect by filling recently fired pitharia with water, and leaving them to stand for several days (Jones 1986).

The use of insecticides in storage

> " leaves of a certain plant were pounded and soaked in water in a pot. The leaves were left to ferment for three days. Blood was added...the herbal medicine (yath)...was sprinkled on leaves...stem borers (nyanginja) died within seven days leaving the crop clean."
>
> (Chitere & Omolo 1993).

Traditional ways of storage include the use of different substances in the storerooms against insects and other pests. The remedies are mainly used to deter pests in the storerooms, and only in a few cases were they applied against field infestation. The use of plants or parts of plants, plant extracts, oils, ashes or minerals, is recorded in the ethnographic literature from various parts of the world (Golob & Webley 1980). Many natural plant substances have some chemical effects on insects (Crainge & Ahmed 1981), and they are usually processed by chopping or powdering to release the agents. At the present day, the most frequently used plants are *Azadirachtha indica* (neem) and *Acorus calamus* (sweet flag) (El-Nahal *et al.* 1989), particularly in the Indian subcontinent. Other plants still in use are Capsicum peppers and chillies, mainly in Nigeria, Gambia, Sierra Leone, Malaysia, the Philippines, and Mexico. *Artemisia* spp. (wormwood) are used in Pakistan, and *Trigonella foenum-graecum* (fenugreek), in India. Similar insecticidal properties are present in the bark of *Eucalyptus* spp., *Cinnamonum zeylanicum* (cinnamon), *Melia azedarach* (Persian lilac, chinaberry), and also plants like *Annona reticulata* (custard apple), *Mentha spicata* (spearmint), and *Piper nigrum* (black pepper) (Golob & Webley 1980).

Plant oils are also traditionally used against insects (Regnault *et al.* 1993). Tests on neem (*Azadirachtha indica*), sweet flag (*Acorus calamus*) and Malabar nut tree (*Justicia adhatoda*) have shown very good control of different species of insect pests (Golob & Webley 1980; Don-Pedro 1987). Vegetable oils are very effective in controlling the reproduction of Bruchids in particular (Schoonhoven 1978; Mensina & Renwick 1983; Pereira 1983), although similar substances have been advocated against both *Rhyzopertha dominica* and *Sitophilus oryzae* (Sighamony *et al.* 1986). As Bruchids lay their eggs on the outside of pulses, probably their effectiveness depends on the absorbing properties of the bean surface, impeding oviposition by the beetles (Delobel & Malonga 1987).

Experiments with animal and plant oils, such as groundnut, castor, palm kernel, maize, cotton seed, Soya bean and coconut oil, mustard, gingelly and sunflower, sesame, shark liver, safflower, and also almond oil have shown that they can be very effective against insect populations (Golob & Webley 1980). However, vegetable oils sometimes leave a taste on the treated product, which may become not only characteristic of a particular storage technique, but also the derived flavour may become a desired characteristic for the area that practices the method. Coriander may provide an example of such a use, but, also traditionally, animal fat, as well as cashew nut oil, castor, garlic, neem oils, and sour orange and citrus fruit peel extracts, as well as lemon, grapefruit, lime and hardwood tar oils, were used. All may have been seen as enhancing the taste of the grain or pulse, or at least masking the taste of any infestation or mould.

The use of ash, sand, mineral powders and even chaff or seeds in large quantities, may have aimed to fill up the voids in grain bulks to provide a barrier to insect movement. Sand particles, ash and mineral powders also destroy insects by abrading their integument and reducing their ability to resist dehydration (Alexander *et al.* 1944; Miller 1987). In addition, wood ash, plant ash, used extensively in East Africa (Miller 1987), and cowdung are very effective against insects, and are used traditionally in different parts of the world. The use of minerals like kaolin, mercury, lime, various soils as well as common salt and boric acid is also very widespread. Like ash, they have absorptive and abrasive properties and, as a result, are insecticidal, or simply act as deterrents (Golob & Webley 1980).

The ethnography of Greek storage

Different storage techniques

Storage in Greece has been practised since before the beginnings of settled agriculture (Rowley-Conwy & Zvelebil 1989), and the remnants of the ancient methods can still be found in remote villages or on small islands. Different storage methods were preferred in different parts of the country. In northern Greece, in Thrace, storage was in the front wall of the house, or the chimney wall. Called "αμπάρι", the container was made of cypress wood, 7 to 8 cm thick, divided in different compartments for different kinds of grain. In west Thrace on the other hand, the "πισινό" was utilised, a storeroom in the back part of the house (Megas 1968-9; 39, 47).

In Thessaly, storage was in "γιαπιά", compartments constructed of unbaked bricks. In earlier times, they were made in the main room of the house, but now they are outside. However "γιαπιά" are very liable to mould and insect infestation, perhaps because of their removal from the dry atmosphere of the house. Another way of storage practised in Thessaly, as well as in Aetolia and Macedonia, is storage in κοφίνες, large wicker baskets 1.5-2.5 m in height, and 1.60 m in diameter. They may have a capacity of up to one thousand kilos of grain, and were often placed on a kind of elevated platform (Megas 1968-9, 8-9). The grain was covered with straw, and sometimes a thick layer of mud (*idem*), or cow dung was put on the top (Halstead, in Wardle 1987).

On the island of Lemnos, in the northern Aegean, completely different methods of storage are practised in different parts of the island. One method is to store the grain in the yard of the house under a pile of straw. People also use fig wood to cover the base of the pile, and mix strong smelling herbs with the grains as a precaution against insect attack. Elsewhere on the island they make "γούβες", pits with the shape of large jars, 2-3 m deep. A stone plate, lined with clay, is used to hermetically close the pits (Megas 1968-9, 68). In the western part of the island, people store grain in a large room of the house, next to the stable. The storage compartment is separated from the main room with a wall, 1.5 m high. It is also divided with partitions made of unbaked bricks for different kinds of foodstuffs (*idem* 1968-9, 72).

Another method of storage was in hollowed out tree trunks, which the inhabitants of the island call "διψέλλα". An example of this technique, dated to the nineteenth century, can be found in the museum on Chios. It is 1.5 m high and 1.2 m in diameter, hollow inside (Triantafyllidou-Baladie 1979, 154).

In the Dodecanese, as, most of the time, the house consisted of a single large room for all the family, grain was stored in a wooden case, kept inside the room and used also as a bench, called "πάγκος". For large quantities of grain, the inhabitants used an elongated pit, usually opposite the chimney wall, 1.2 m deep and covered with wooden planks; this was called "ταβλαντό" (Megas 1968-9, 207, 215).

In Rhodes, the farmers store in pits, which they cover first inside with red earth; afterwards they burn pieces of wood inside the pits for 24 hours. In this way the walls of the pits become impenetrable to the insects (*idem,* 207).

The traditional use of insecticides in Greece

Different plants and other substances are used traditionally in Greece against insects. Dill, *Inula graveolans*, and marjoram, *Origanum vulgare* are mixed with wheat against *Sitophilus granarius*. Also leaves of the introduced tree *Eucalyptus* sp., and those of fig, bay and osier are used in the crop against insects and other pests. Bay leaves are a very widely used insecticidal substance in Greece (Saim & Meloan 1986), and they are even put in wardrobes against moths, as well as in larders, etc. Pulses are often soaked in olive oil before storage (Halstead 1990), and kaolin or lime is reported to be mixed with cereals against *Sitophilus granarius* (Buchelos & Mourikis 1995; Golob & Webb 1980). Honey with petroleum jelly is used to coat the windows of tobacco warehouses, in order to trap *Lasioderma serricorne*, the tobacco beetle. Fires are also lit around infested fields and warehouses against insect infestation (Mourikis in Golob & Webb 1980). Cowdung, sand, burnt wood are often mixed with cereals in different parts of Greece, and dung-lined baskets are known from rural Greece (Halstead 1987). Salt remains the most common preservative for meat and fish (Gousgouni, pers. comm.).

Storage on Santorini

In Akrotiri, Theran traditional farmers consider lack of rain and insect pests as the most important enemies of their yearly crops. The σκουλήκι worm, or else the ψείρα louse in the wheat, and the πουλί bird in barley and pulses might easily destroy a year's struggle. Large earthen jars "κουρούπια" were used till recently for storage. Most of the time they were painted on the inside, so as the inner surface would be "as solid as glass". Usually they were covered with sacking (λινάτσα), and on the top of it was put limestone earth or limestone earth mixed with chaff. Sometimes, farmers would put an iron rod in with pulses, perhaps aiming to keep the temperature of the seeds down.

Other farmers had wooden chests for storing large quantities of grain (2-3 thousand kilos). When they needed more receptacles, they were using linen sacks which were afterwards covered with a large quantity of chaff. Another secure way of storage was the "χαβούζες", pits which, after being filled with grain, were covered with a layer of chaff, sacking, and lastly a layer of earth. The grain was taken out of the "χαβούζες" with the use of wooden buckets, "κάδοι", of half a "κιλό", what was called kilo was in fact a weight measure of around 29 to 30 kilos. In addition, there were the "υπόσκαφα", storerooms inside caves on a dry rocky part of the mountain. The caves used were either artificial or natural, and their mouths were closed by wooden doors. During the Second World War German occupation, the farmers used to hide their crops in "υπόσκαφα", as the occupying troops confiscated all food. The villagers in Akrotiri stored in their houses food needed for short periods, whilst larger amounts of cereals, pulses and other foodstuffs were stored in "υπόσκαφα" for more security and better preservation. Barley and wheat, according to the farmers, could be kept from two to three years, while pulses would last for ten. These cavities were also used during the last eruption of the volcano in 1926 as shelters for the people of Santorini (Anon 1945). Stories are still told about families that were found dead from suffocation in such caves. "Υπόσκαφα" are still used as barns, or when near the sea, as a kind of small houses for the boats (see also, Shaw 1990).

Pulses and cereals were stored in the same way. For large scale storage (χαβούζες, κασσέλες), crops from different years were mixed in the same receptacle. In contrast, in pithoi, the older seeds were put nearer to the top and were separated with sacking from the new crop, the idea being to consume first the older crop before it became unfit for eating. The villagers claim that two to three years old vetch is better for cooking than fresh, which is very hard. Vetch, before being stored, was always dried in the sun and sieved thoroughly. When already evidently infested the seeds were floated in water and the "attacked seeds" (χτυπημένοι σπόροι) were easily separated and thrown away, or used for feeding the animals. Soaking of many legume crops was also necessary to remove natural organic chemicals poisonous to humans.

The archaeology of storage

There has been extended discussion about the origins of food storage; some believe that it was a well known practice before the beginnings of agriculture (e.g. Sahlins 1972, 31-32), while others regard it as a result of the new economic conditions that came with agriculture (Flannery 1971; Harlan 1975). One way or the other, it seems that storage of crops was practised since very early times. The most ancient method

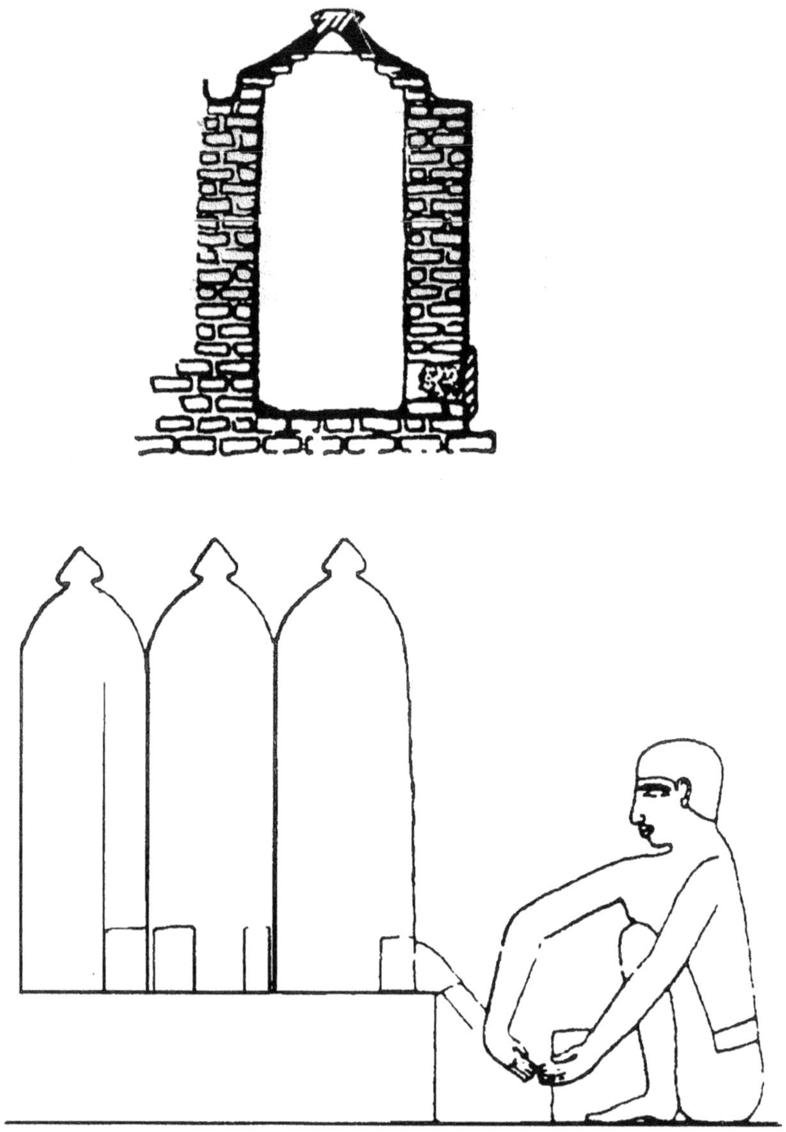

Fig. 5.3. Upper : Section of a brick granary in Saqqara tomb 3038, Egypt (after Spencer 1982). Lower : Detail of a wall painting showing a 3-domed storage chamber, Tomb 50 (Ti), Saqqara (5th Dynasty) (after Baines & Malek 1980).

used for preserving goods is likely to have been cave storage. Caves were occupied by the middle Paleolithic, and as far as storage is concerned, they maintain a steady temperature and a secure environment for preserving food commodities; they were used until recently in the Mediterranean region. Pit storage is also a very well known method worldwide, recorded almost everywhere in Europe, Northern Africa, Asia, from Turkey to India and China, North and South America (Sigaut 1988). It is still used around the Mediterranean and in North Africa.

Storage in Egypt: the pictorial and archaeological evidence

A number of different types of granary were used by the ancient Egyptians, and there is evidence for communal storage from the Early Dynastic period onwards (cf. Mills 1992). The basic type, the *shouna,* was tower or dome shaped, and may have been a descendant of the mud huts used around the Nile's Delta as shelters. The granaries were ovoid, had an entrance in the roof and a ladder to reach it. Dome-shaped storage chambers, up to 5 m high and 2 m in base diameter, were built in one or more rows, usually situated within an enclosure. All appear to have a roof opening and a shutter above or near the base (Fig. 5.3). The roof opening was used to fill the storerooms up while the opening underneath was to take out the grain. Small openings instead of doors served to stop larger pests from entering the storeroom (Levinson & Levinson 1985; Strouhal 1996, 101).

Another type of granary was used at the end of the Old Kingdom period; it consisted of a series of storerooms, with an opening and a shutter, all forming part of a single flat roofed building. During the New Kingdom, this type of granary declined and dome shaped granaries came into use once more (Levinson & Levinson 1985). Storage also took place in a type of protected yard, in the open. The exposed heaps of grain formed a very thick crust which secured them from insects and other pests (*idem*). In some cases, huge silos were used, for example in the Middle Kingdom at El Amarna (Borchardt 1916).

Greek storerooms

"'Αποθήκη, καλιά, πιθεών, πύργος, ρόγος, σιτοβολεών, ταμεῖον", are terms used during Classical times for places that were associated with grain. The names (except ρόγος and σιτοβολεών) are mostly used to describe in general a place where the provisions are kept. There appear to be no specific names for different types of storerooms and storage (Amouretti 1979, 60).

Storerooms in Greece during early antiquity were either collective, or exclusively for the needs of a small household. The Cretan square subterranean magazines are an example of large scale storage for the needs of the palace; 20-30% of the ground floor areas of the early palaces at Knossos, Mallia and Phaistos have been interpreted as used for storage (Halstead 1981). Later, during the fourteenth century BC, when the storage capacity of the palaces appears to have been reduced, Linear B tablets (Fig. 5.4) refer to large quantities of grain from Dawo, an area located somewhere around the Messara plain (Halstead 1992).

Large quantities of grain were kept in huge pithoi in Crete. Different sizes of vessels and receptacles made of other materials, wicker, wood, were also used for storage. The magazines of the West Palace section at Knossos provide an example of the various different types of storage used during Middle Minoan II at Knossos. Pithoi during MM II were replaced by cists or "κασσέλες", sunk into the floors, ninety three in all. Their walls lined with tin, they were used for storing valuable objects and not necessarily food. A series of more square and more capacious receptacles made of limestone (and not of gypsum slabs as in the case of κασσέλες) were also found in the so-called Long Gallery. Oblong cavities about six inches deep were cut in the base blocks of these receptacles. The whole of their interior surface was covered with lime mortar and most probably they were used for the storage of liquids (Evans 1935, IV, 630-632), although the treatment is reminiscent of that recommended for grain stores by the Classical writers. Begg (1975) gives an account of Minoan storerooms during the Late Bronze Age.

At Assiros in Macedonia, during the Bronze Age, there is convincing evidence for crop storage, and the floors of several crop storerooms, destroyed by a fire *ca.* 1350 BC, have been excavated. The crops were put in pithoi, clay bins and perhaps sacks or other perishable containers that have not survived. In some cases, the pithoi were placed inside pits (Jones *et al.* 1986).

An interesting point about storage in Antiquity, is the capacity of the large jars which were used for storing the various commodities. Measurements have shown that the capacity of pithoi was from 142 lt. (h: 0,92 m) to 1085 lt. (h: 1,74 m) at Mallia, while the ones found at Olynthos XII were about 170 lt. (h: 0.94 m) (Amouretti 1979).

a.

b.

c

d

Fig. 5. 4 : Linear B signs relating to cereals and storage : a. ears of barley. b. 'bifid' cereal sign. c. 'granary' signs. d. elongate tablet of 'cereal' class (Handlist no. 13) (after Evans 1935).

Roman granaries

Elaborate storage facilities for foodstuffs was Rome's answer to the proper organisation of the food supply for both cities and military units. Secure buildings, well placed and easy of access, with space for loading and unloading are widely recorded (Rickman 1971). The normal Roman granaries were long, narrow, rectangular buildings, strongly constructed, elevated from the ground and well ventilated (Fig. 5.5). *Horrea*, as the Roman granaries were called, often had huge storage capacity, and Ostia, the port of Rome offers good examples of such constructions. In Ostia, at least three of the great *Horrea* had raised floors, to protect grain from damp and overheating; air tunnels were created underneath the floors. The methods of storing grain were in sacks, bins, or in large heaps in the floor. However, in Ostia, divisions inside the rooms served a purpose similar to that of bins. Rickman (1971, 86) believes that in Ostia grain was probably stored in sacks.

Underground storage in antiquity

> "...who had a ... in his courtyard; and they went down to it. And the woman took and spread a covering over the mouth of the ..., and she scattered grain ... upon it; and nothing was known of it."
>
> 2 Sam 17:15-20

Research on prehistoric pit storage has been undertaken in Great Britain, and both Neolithic and Iron Age pits have been recorded and discussed (Field *et al.* 1964; Piggott 1958). Reynolds (1974; 1979) has experimented with the technique, although without insect infestation. Sumero-akkadian literature gives information about pit storage, and Graeco-Latin literature acknowledges pit storage as a sound technique practised in Spain, Carthage, Thrace, and Cappadocia (above). Pit storage in Byzantium provided food for Caesar's army (*Bel. Afr.*, 65) and ancient Hebrew, Aramaic and Ugaritic texts allude to pit storage, although there appears to be no specific term used to describe the method (Currid & Navon 1989).

Pits, probably for storage have been excavated in Palestine, at the site of Ain Mallaha (Einan) as early as the Epipalaeolithic (Perrot 1957, 91-110; 1960, 14-22), and there are numerous later examples; pit storage appears in all archaeological periods thereafter in most Palestinian sites. The tradition of storing in subterranean silos is also very old. There are several examples, like the silos in Sicily that Diodorus Siculus records (XIII, 83). Other underground silos full of jars have been excavated in Troy, Pompeii and Ostia (Mauny 1979).

Fig. 5. 5 : Elevated floor of Roman granary at Corbridge, Northumberland, England

6. The Geography, History and Archaeology of Thera

Introduction

The small village of Akrotiri, the Greek word for cape or headland, is situated in the southern part of the island of Thera. To the south of the modern village, closer to the sea lies the prehistoric site of Akrotiri. Thera, or Santorini, is a volcanic island, situated in the southern Aegean. Its present form is the product of at least ten major eruptions over the past one million years (Pichler & Friedrich 1980, 15). Where today there is a group of islands, Therasia, Aspronisi, Palaea Kameni, Nea Kameni and Santorini, there was once one island, Stronghule. In Greek, "στρογγυλὸς" means round, and Stronghule, the round one, was, in Antiquity, a typical name for a volcanic island as, for example, Stromboli in Italy. Thera was completely destroyed by a Plinian eruption at the beginning of the Late Bronze Age (Doumas 1983). Prior to the eruption, Thera is thought to have had a water filled caldera in its central part (Friedrich *et al.* 1980; Ericksen *et al.* 1990). Pichler and Friedrich (1980) believe that, before the Late Minoan destruction, there was a central cone about 500-600m high on the island, although Hedervari (1978) suggests a cone 1500m high, and Heiken and McCoy (1990) argue for a large number of small cones.

The mountains of Profitis Ilias, the highest peak of the island, about 500m high, Mesa Vouno and Mikros Profitis Ilias all existed in Minoan times, while, in other parts of the island the explosion and the deposition of thick pumice layers during the eruption has changed the landscape completely (Hardy & Aston 1990). The surface of Thera today, together with the islands of Therasia, Nea and Palaea Kameni and Aspronisi, covers about $90m^2$. Having in mind the collapse of the northern caldera, and also the pumice generated area south-east of Profitis Ilias, we can assume that Late Bronze Age Thera was about $103m^2$ in extent (*idem.*).

The processes of destruction caused by the Plinian eruption could have lasted from a few hours to some weeks (Pichler & Friedrich 1980, 16-23). The eruption was preceded by an earthquake and, between the two events, there may have been a gap of a few months or several years (Pichler & Friedrich 1980), although Davidson (1978, 285) believes that there is no basis for a gap.

Thera -the mythology

The first colonisation

The volcanic genesis of Thera was the inspiration for the myth of Euphemus, one of the Argonauts. Euphemus following Triton's advice, threw a lump of earth into the sea, north of Lake Tritonis (in Tunisia) and it was transformed into the island of Thera (Doumas 1983, 20). Herodotus (IV, 145ff) talks about Kalliste - Greek for "the most beautiful", the first name of the island- and its colonisation by Kadmos the Phoenician. The story is told in connection with the return of the Minyans, the descendants of the Argonauts, to Lacedaemon, when they were driven from Lemnos by the Pelasgi, in the third generation from the golden fleece mission (Lambrou-Phillipson 1990). Kadmos, the son of Agenor, a king of Tyre (or Sidon) in Phoenicia, while looking for his sister Europa, stopped at Thera and for some reason left there a number of "Phoenicians" and with them Mebliarus. After eight generations of the descendants inhabiting Kalliste, Theras, the son of Autesion and Kadmeian by birth, left Sparta where he had been brought up, when he came of age, because he did not want to live under the authority of others. He took with him men of each of the tribes of Sparta, as well as Minyans and went to Kalliste with three 30-oared ships. He settled there and renamed the island Thera. Thucydides also talks about the Carians and "Phoenicians" colonising the Aegean islands (I, 8), and the way Minos and his navy drove them out (I, 4).

It would be unrealistic to infer, based on the stories told by Thucydides and Herodotus, Near Eastern origins for the inhabitants of Thera. However, the myths imply close relations with the Near East, and the significance of the small island. The relations of the inhabitants with Tyre, one of the largest textile centres on the Palestinian coast (Forbes 1956), the "Phoenicians" and the Argonauts, according to the myth, are perhaps reminiscences of a long distance trading connection. The story of Jason and the Golden Fleece has been interpreted as trade expeditions in order to obtain luxury items, maybe silk garments (Scott 1993). Some silkworm species produce bright yellow cocoons rather than white or cream ones, and the silk thread obtained from them could be described as golden.

The Atlantis story

The eruption that destroyed the Late Bronze Age settlement of Akrotiri at its peak was one of the largest in the Mediterranean during the present interglacial, comparable with that of Krakatoa (20 May 1883) in Indonesia (Doumas 1983). All the island must have been covered with pumice and everything was destroyed. There is no way to be sure about the length of the Plinian eruption, but most probably everything happened in a short period of time (Sparks & Wilson 1990).

There has been a long discussion about the destruction of Minoan Crete, and whether it was related to the Santorini eruption (Marinatos 1939). Different theories talk about destruction due to earthquakes, fire, tsunamis, or pirates. The major volcanic catastrophe of Santorini may have been one of the reasons for the decline of the Minoan civilisation. The date of the eruption remains, however, problematic (Baillie 1998; Buckland *et al.* 1997; 1998; Manning 1998; Zielinski & Germani 1998a & b). However, in the literary record a major catastrophe is recorded at around 1500 BC.

In Egypt, during the XVIII Dynasty (after 1500 BC), The Leiden Papyrus records earlier catastrophic events (Admonitions of Ib-Wer 3, 6-9, Papyrus of Leyden, in Doumas 1983, 149):

> "...For nine days there was no exit from the palace and no one could see the face of his fellow...
> It is inconceivable what happened in the land-to its whole extent confusion and terrible noise of tumult: oh that the earth would cease from noise... the towns are destroyed... Upper Egypt suffered devastation... blood everywhere... pestilence throughout the country...
> ...Men no longer sail to Byblos. What shall we do for our mummies...and for oils with which chiefs are embalmed as far as the country of the Kefti? They come no more...
> ...The sun is covered and does not shine to the sight of men. Life is no longer possible when the sun is concealed behind the clouds. Ra has turned his face from mankind. If only it would shine even for one hour! No one knows when it is midday. One's shadow is not discernible. The sun in the heavens resembles the moon..."

More evidence from Egypt has been recently discussed by Davis (1990), including a stele erected at Thebes by Ahmose, the first Pharaoh of the XVIII dynasty. Vandersleyen (1967) translates the text as follows:

> (7) "the gods expressed
> (8) their discontent...The gods (made?) the sky come with a tempest of (rain?); it caused darkness in the Western region; the sky was
> (9) unleashed, without... ...more than the roar of the crowd;...was powerful...on the mountains more than the turbulence of the
> (10) cataract which is at Elephantine. Each house,...each shelter (or each covered place) that they reached...
> (11) were floating in the water like the barks of papyrus (on the outside?) of the royal residence for ...day(s),
> (12) with no one able to light the torch anywhere......."

The catastrophe described in the "Tempest Stele" is placed by experts between 1500-1528 BC or 1539-1517 BC. Dugmore (pers. comm.), however, points out that nowhere does the recorded thickness of tephra from the Santorini eruption in the Eastern Mediterranean support such a dramatic impact.

Plato's story about Atlantis was thought (Galanopoulos 1969; Luce 1969) to be relevant to the Santorini eruption. Solon travelled from Athens to Sais in the Nile Delta, around 600 BC, where the Egyptian priests told him the story of Atlantis. The date given for Atlantis' submergence is 9000 years before Plato, but it has been suggested that the Egyptian symbol for 100 was mistaken for 1000, so maybe the catastrophe occurred 900 years before, at around 1500 BC. Plato in his dialogues *Timaeus* and *Critias* writes the tale as he heard it from his grandfather, Solon. In *Timaeus* he writes:

> "But afterwards there occurred violent earthquakes and floods; and in a single day and night of misfortune all

> your warlike men in a body sank into the earth, and the island of Atlantis in like manner disappeared in the depths of the sea. For which reason the sea in those parts is impassable and impenetrable, because there is a shoal of mud in the way; and this was caused by the subsidence of the island."

Maybe Atlantis never existed. However, the description of the big disaster delivered to Plato by his grandfather may have been nothing other than memories of the Plinian eruption of Santorini.

The chronology of the eruption is still under discussion. The existing literary sources clearly support a late date for the destruction of Thera. Their value as arguments for the chronology of Thera, is questionable. However they do appear far more reliable than the results of radiocarbon dates, which span a considerable period of time (Housley *et al.* 1990).

Climate and vegetation

The Aegean

The area of the Aegean is characterised by a very varied climate. The North is colder than the south, and the west drier than the east (Doumas 1983). The Cyclades (Fig. 6.1), the group of Aegean islands to which Thera belongs,

> "are rockbound and barren, and they consist of plateaux of undulating surface" (Anon. 1945).

Their climate is generally mild, and there are no strong contrasts in temperature, except in the valleys of the larger islands, such as Andros, Naxos and Siros. Rainfall levels are higher during winter, whilst the summer rainfall levels are very low especially during July and August. The period of marked rainfall, (*c.*10 mm per month) is from October to May, and the total rainfall varies from northern islands to southern ones; for example Andros, the most northerly island has 0.62m, while Thera, has only 0.4m (Anon. 1945).

Santorini today - Climate and vegetation

Santorini is a very arid island, one of the driest spots in the southern Aegean. The 30 year mean rainfall is 378 mm p.a. (Marinos & Marinos 1978). However the volcanic soils of Santorini have a high capacity for retaining moisture, a critical factor in the growth of vegetation. The flora of the island is very rich for an island so small. This fact made Rackham (1990, 385) assume that the volcanic eruption did not destroy completely the island. However, the recolonisation would have been easy, since it is geographically well placed to receive propagules.

The wild vegetation of Santorini can be categorised as follows:

> "maquis" (trees or shrubs such as *Quercus coccifera* and *Phillyrea media* and also lentisk *Pistacia lentiscus*).
> "Phrygana" (undershrubs like *Cistus* and *Salvia*)

and

> "steppe".

The trees that grow there today are prickly-oak (*Quercus coccifera*), juniper (*Juniperus phoenicaea*), wild pear (*Pyrus amygdaliformis*), carob (*Ceratonia siliqua*) and also pines and cypresses.

Past climate and vegetation

McCoy (1980, 96), based on terrestrial and oceanic isotope measurements, claims that there were no significant climatic changes in the southern Aegean during the Late Bronze Age, although he believes that the air temperature may have been warmer by 1^0 C around 3500 BP. According to Rackham (1990), the climate has not changed at all for the last 2500 years. The main palynological

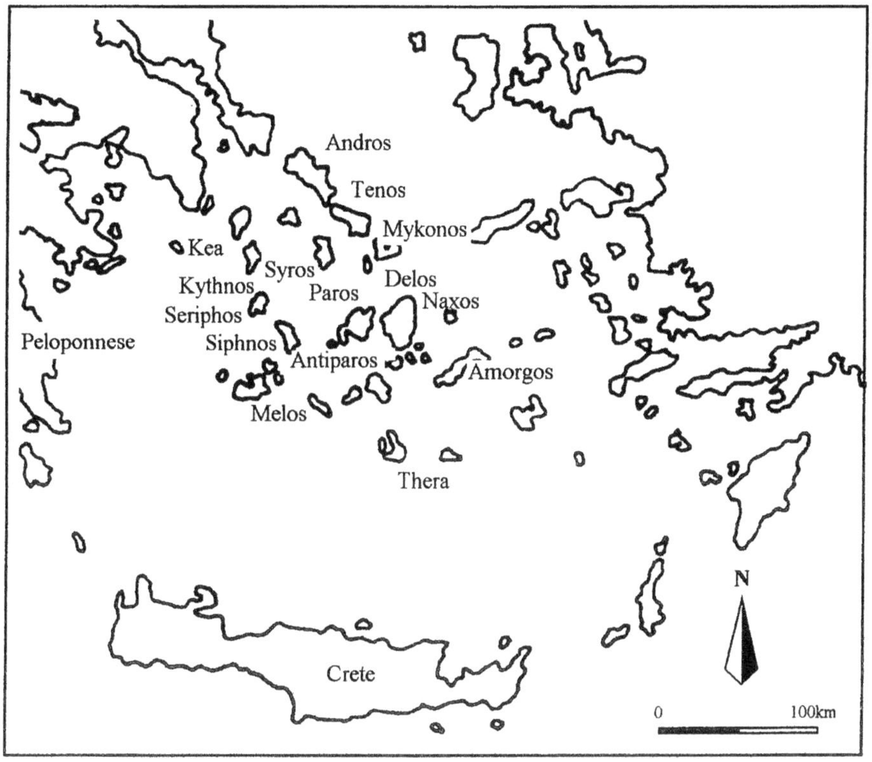

Fig. 6. 1. The Cyclades and southern Aegean Sea

work in the southern Aegean has been undertaken by Bottema (1980, 211) in Crete. The conclusion of this research is that Crete,

> "was dominated by deciduous broad-leaved species, especially oak", before forest clearance.

On Santorini itself, the reconstruction of past climate and vegetation is problematic. Whilst the preservation in the settlement is amazing and even three storey buildings have been recovered, the nature of the site, as well as the way it is preserved does not allow study of the surrounding contemporary environment. As the island is covered with thick pumice layers, trenches have not been cut outside the inhabited area and environmental samples have not been taken. Rackham (1990), based on the available information, claims that the island must have been a little less arid than it is today. However, even if the orography of Santorini was only 200m more than the present one, its climate would have been completely different, and it would have been capable of supporting far more woodland, as the mountain peak would attract the clouds and therefore the rainfall would have been higher. If we accept the height of the cone suggested by Hedervari (1978), 1500m, Santorini would have been more like Crete in terms of climate. In addition, Limbrey (1990) believes that Bronze Age soils, although very sandy, had a very low clay content, and therefore were not as poor as previously thought (Davidson 1978). It is quite possible that Late Bronze Age Santorini had patches of lush vegetation or even extensive and luxuriant woods of many trees, as, for example, on present day Samos, with its peak of 1153m.

Pleistocene plant remains from Thera and Therasia include the palm, *Phoenix*, olive, *Olea*, *Pistacia* and *Tamarix* (Friedrich 1978; Venitzelos 1990). The excavations on the Bronze Age site have given some indications of the vegetation. An "oak stump" was found in the first excavation of Akrotiri and "branches of vine and olive" have been recorded from Therasia (Mamet 1874).

Settlement pattern and land use

It is difficult to discuss prehistoric settlement pattern in Santorini. Usually, knowledge of settlements is based on buildings, walls, burials and pottery found under the mantle of overlying pumice (Hardy & Aston 1990). However, approximately 43 km^2 of the island were lost in the collapse of the caldera and only a small part of the original island has survived, with most of the remainder being covered with very thick pumice. As a result, we can only make assumptions about the density of settlements. Hardy & Aston (1990), comparing the settlement patterns of the island with the ones on Melos (Renfrew & Wagstaff 1982, 22: 20 Bronze Age sites on an area of 33km^2), think possible the existence of 62 sites on the island, of which around 32 might have been preserved. On Classical Melos, 59 people per km^2 could be supported. (Wagstaff & Cherry 1982, 145), and this number was higher during the nineteenth and the twentieth centuries. In 1838, Thera was inhabited by 39 people per square mile and, before World War II, by 139 people per km^2. At present, we know of seventeen Late Bronze Age sites which have been found in the south-west part of the island (Sarpaki 1987) (Fig. 6.2). Six of the above sites are located near Akrotiri, but probably this merely reflects the fact that more surveys have been undertaken around this area (Sarpaki 1987). Doumas (1983, 129) believes that the south-west of Thera was more densely populated than the North. It should be stressed, that, despite all the estimates mentioned above, without extensive survey being undertaken on the island, only guesses can be made about the settlement patterns.

Hardy & Aston (1990) also suggest that the most important port of Thera, and the major settlement, might have been located under Kamari and near Emporion, as the sheltered bay into the caldera would have been an ideal harbour. However, it is a commonplace among the islanders today, that Akrotiri is the most protected place of the island and most of the fishing boats find shelter in its harbour.

Crop husbandry - present

On Thera, the main crops cultivated for domestic use are barley, *Hordeum vulgare*, wheat, *Triticum sp.*, peas, *Pisum sativum*, Spanish vetchling, *Lathyrus clymenum*, sesame, *Sesamum indicum* L., millet, *Panicum* sp., a fodder pulse crop, *Vicia articulata*, olives, *Olea europea*, fig trees, *Ficus carica*, grapes, *Vitis vinifera*, bitter vetch, *Vicia ervilia*, cotton, *Gossypium* sp., (probably *arboreum*) and beans, *Vicia faba*; the last three species are no longer cultivated on the island. Sometime before the Second World War, tomatoes were introduced and took over a part of the cultivation, and now redundant tomato factories still exist in Perivolos and Vlyhada.

As there is a lack of water, dry farming and intercropping are practised (Sarpaki 1987). *Lathyrus clymenum* is drought resistant and is the most rewarding crop for the volcanic soils of the island. Careful choice of the variety of crops to be sown, and diversification, as well as small fields, are the answers of the inhabitants of the island to the risks of production loss (Sarpaki 1987).

Crop husbandry - past

The archaeobotanical work undertaken on the site by Sarpaki (1990) revealed the cultivation of a wide range of pulses: *Lathyrus clymenum, Lens culinaris* and *Pisum sativum*, *Lathyrus cicera/sativus, Vicia ervilia* and *Lupinus* cf. *albus*. Barley (*Hordeum* sp.) was also cultivated. Processing of crops was also going on at the site, and the botanical finds indicate that flour and bulgur were made at a household level.

Crops cultivated and crop husbandry practices are very similar between the present century and the Late Bronze Age, and Sarpaki (1990) talks about small fields around the Late Bronze Age settlement, cultivation in terraces, and drought resistant crops.

Old excavations

The earliest excavations took place at Balos, directed by Mamet and Gorceix (Fouqué 1879, 118-120; Perrot & Chipiez 1894, 150). Walls of dwellings were found. The stones of the walls were of cut and laid ashlar blocks, and the inner sides of the walls were plastered and painted yellow. The structures were interpreted as service rooms as they had no windows and the floors were of black soil. Four of the rooms were used for storage of crops (lentils, barley, peas) and chaff, while another had been used as a workshop. Obsidian and pottery were also found in abundance (Page 1970). Remains in Alaphouzos Quarry in Therasia were also excavated, first in 1866 by Alafouzos and Nomikos, then by Fouqué in 1867, and in 1870 by the French School at Athens under the supervision of Mamet and Gorceix (Fouqué 1879, 96-106; Perrot & Chipiez 1894; Page 1970). In 1867, Fouqué excavated some of the ravines south of Akrotiri, and in 1870 Mamet and Gorceix excavated several other parts of the area. However, the French did not publish precise information about the sites excavated, and they have not been re-discovered (Page 1970). Zahn excavated in 1899, 600m away from the present site of Akrotiri, to the east, between the present day excavation and Potamos valley. According to Doumas (1974), this area was also a part of the town. and he has estimated the area of the Akrotiri settlement at 200,000m^2. Of this, only 10,000m^2. have been excavated. He also claims that the Balos and Mavromati quarry sites were suburbs of the town (Doumas 1974).

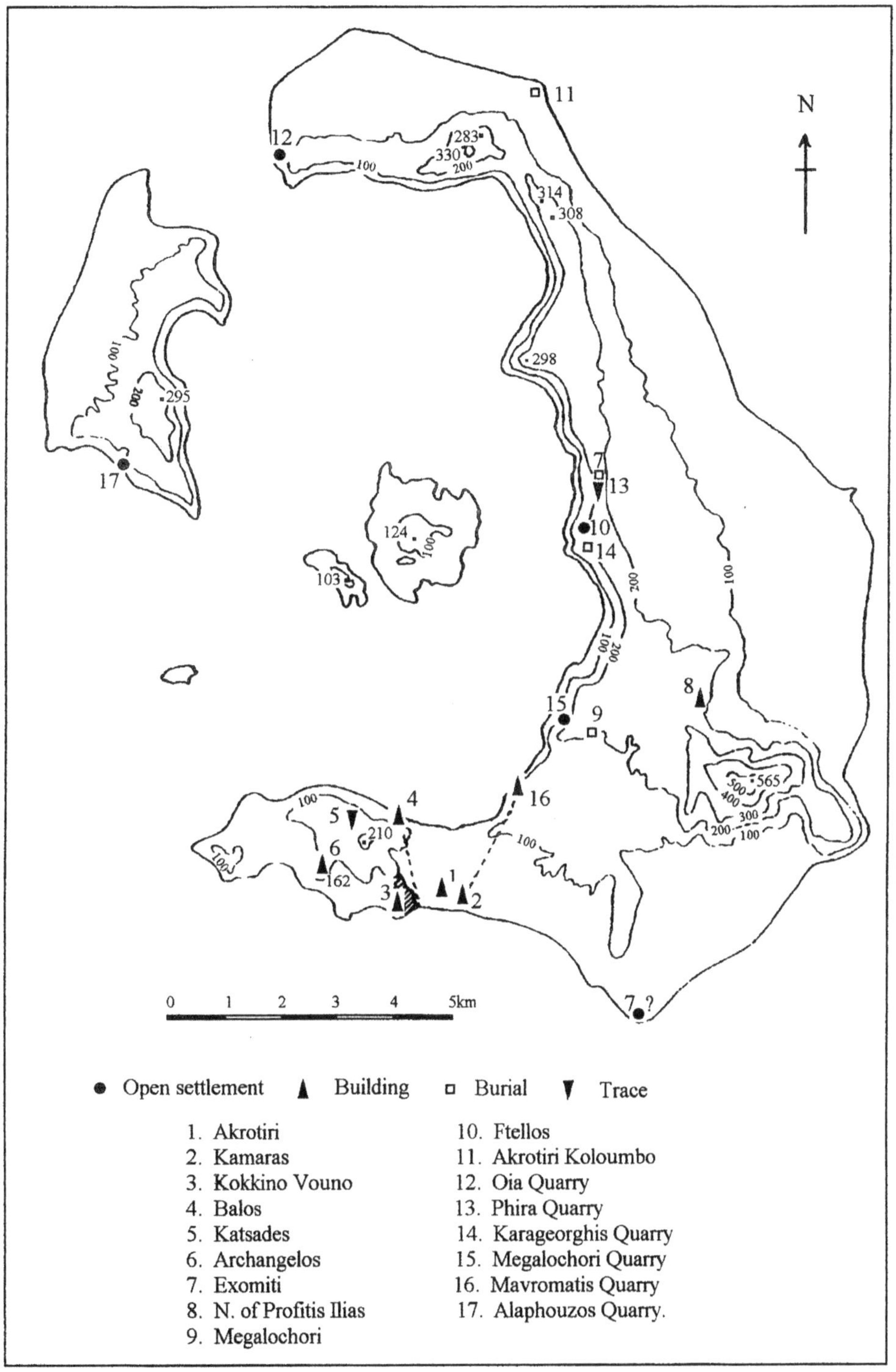

Fig. 6. 2 : Bronze Age sites on Thera (after Sarpaki 1987).

New excavations (Fig. 6.3)

Marinatos (1939) related the decline of Minoan civilisation around 1500 BC, to the destruction of Thera. In 1967, he started excavating the site of Akrotiri, and he continued until 1974, when he died while supervising the excavation, and C. Doumas undertook the directorship. Excavations have continued, but on a far smaller scale. Doumas hired specialists from different disciplines, and the understanding of every aspect of the life of the settlement has become a priority; this approach made the palaeoentomological study possible.

Description of the site

Akrotiri must have been inhabited since the Neolithic, but no detailed information about this phase of occupation is available, as only sherds of pottery have been found (Doumas 1983). An outline of the Early Bronze Age occupation is given by Sotirakopoulou (1986). More information is available about the Middle Bronze Age occupation of the site, and, based on the archaeological evidence, Doumas (1974) believes that there was a smooth transition from the Early Bronze Age to the Middle Bronze Age settlement. Contacts with Crete are certain, as the pottery known as Kamares ware has been found (Doumas 1978, 782; Doumas 1983, 42). The town was first destroyed during this period by earthquakes, but it was almost immediately rebuilt (Palyvou 1984; 1986; Marthari 1984); the town plan appears to have been very similar to the Late Bronze Age one (Doumas 1983, 43). As a result of rebuilding after the earthquakes, the ground floors of the houses at Akrotiri are below the street levels (e.g. the ground floor of the West House). Before the final destruction of the site, earthquakes caused severe damage twice. The first time the damage was on a small scale, and rebuilding of some houses occurred, as well as rearrangements of upper floors (Palyvou 1984). The second phase of earthquakes was far more destructive, and the town was rebuilt on an even larger scale.

The way the streets were structured (Palyvou 1986), as well as the sewer system, testify to the high degree of organisation (Sarpaki 1987). It has been claimed that the part of the town recovered is the wealthy one (Doumas 1983). The absence of palatial buildings, however, could imply a different social organisation, compared with Crete and its Minoan Palaces. The area excavated, is only a small part of the town of Akrotiri, and does not really allow assumptions about the administration of the settlement. It is possible to divide the buildings recovered into private (areas Beta, Gamma, Delta, the House of the Ladies, the West House) and public (Xeste 2, 3), based on their construction and the archaeological finds (Sarpaki 1987).

Chronology

There has been a long debate about the date of the Theran eruption. As Akrotiri is a Late Bronze Age settlement, a date for Thera is very important since it effectively also gives a date for the beginning of the Late Bronze Age in the Aegean. Until recently, it was believed that the eruption of the volcano occurred at around 1550-1500 BC. Warren (1990a) summarises evidence that supports the absolute chronology, and concludes that we should seek a date well before the end of Late Minoan IA, (1600/1580-1504/1480); according to Manning (1990) the date could be between 1625-1525 BC. The ceramic evidence is not at all precise, and only suggests a wide chronological range, usually merely a *terminus ante* or *post quem* (Manning 1990).

Different methods have been employed so that a date could be established. A number of radiocarbon dates on short lived samples have yielded inconsistent results. The first University of Pennsylvania radiocarbon dates supported a date of *ca.* 1625 BC (Michael 1976; Betancourt & Weinstein 1976), but the radiocarbon dates from Simon Fraser University are more confusing. Seventy separate AMS determinations on forty seeds show a wider scatter than the original measurements and various interpretations are possible (Nelson *et al.* 1990). Other calibrated dates from Akrotiri by AMS (Housley *et al.* 1990) show two ranges, one of 1674-1606 and another of 1554-1534 cal BC, the first one being the most probable.

There is no direct dendrochronological data for the eruption and the data are ambiguous. Baillie & Munro (1988; Baillie 1990) has suggested that a period of reduced growth in the Irish tree ring sequence can be equated with the eruption of the Aegean island of Santorini in the second millennium BC. However, analysis has shown that the 1628 BC event in the Irish tree record need not be volcanic (Buckland *et al.* 1998). An Anatolian tree-ring sequence discussed by Kuniholm *et al.* (1996), shows an extreme growth of tree rings dated by cross-matching with the Irish sequence to the year 1628 BC. Related phenomena have been observed in tree ring sequences from California. However, as we do not have a complete record world-wide of prehistoric eruptions, there are no grounds to attribute the above event to the Santorini eruption, rather than another unknown source (Buckland *et al.* 1998).

Ice core acidity data are also problematic. Explosive eruptions inject volumes of particulate matter into the atmosphere. The bulk of these are either of silicate ash or sulphur compounds, and under suitable conditions acid fallout may be preserved in ice cores (Pyle 1989). However Zielinski and others (1994) have shown that high SO^4 does not necessarily correspond with high volcanic explosivity (VEI) (<4), for example, the 1883 eruption of Krakatoa at VEI 5 had SO^4 residuals calculated to 46 while the Hekla eruption (VEI: 3), although of low explosivity, was high in SO^4 residuals (SO^4: 83). The correlation between Santorini and the ice cores receives further damage from Pyle's (1990) comment that the eruption was likely to have been low in sulphur. Without preservation of tephra from the actual eruption in the ice cores (Zielinski & Germani 1998a & b), any correlation has to be purely speculative.

Population estimates

The fact that the settlement has only partly been uncovered is a major problem in any effort to estimate population numbers at the settlement of Akrotiri. An area of 10.000m^2 has been excavated (Doumas 1983, 45) and ,as the uncovered area is only as estimated one twentieth of the site, it is uncertain whether it is representative of the town as a whole.

The population estimates for the West House (83.3m^2 first floor -146.95 ground floor + first floor), according to Narrol's model (1962, 587; 10m^2/person), are from eight to fourteen people, according to the 16m^2/person model, five to nine people, while Sumner's (1979, 165; 50-100m^2/person) estimate of one to two people seems completely unrealistic (Table 6.2). If the West House was inhabited by five people, 100 stremmata (1000m^2) of land, the labour of three persons and a food storage capacity of 2290kg. would have been required (Narroll 1962). If we apply the eight people estimate, we should accept that five of them were involved in agriculture, and the land needed was 160 stremmata (=1600m^2) and the yearly storage required 3664kg (Sarpaki 1987).

	Narrol's estimate	16m^2/person	Sumner's estimate
first floor (83.3m^2)	8	5 (2)	c.1
ground & first floor 146.95m^2)	14	9	c.2

Table 6-1 : Population estimate of the West House based on the area of the House after Sarpaki (1987).

If the area occupied by the site originally was 200.000m^2, the built up floor area would have been 94260m^2 (Sarpaki 1987) and if multiplied by two (for two storeys, as most of the houses had two or even three storeys), 188520m^2. The calculation (Table 6.3) is based on the area sampled by Sarpaki (1987) (1725m^2 sampled of which 813m^2 was built area and 912m^2 open spaces).

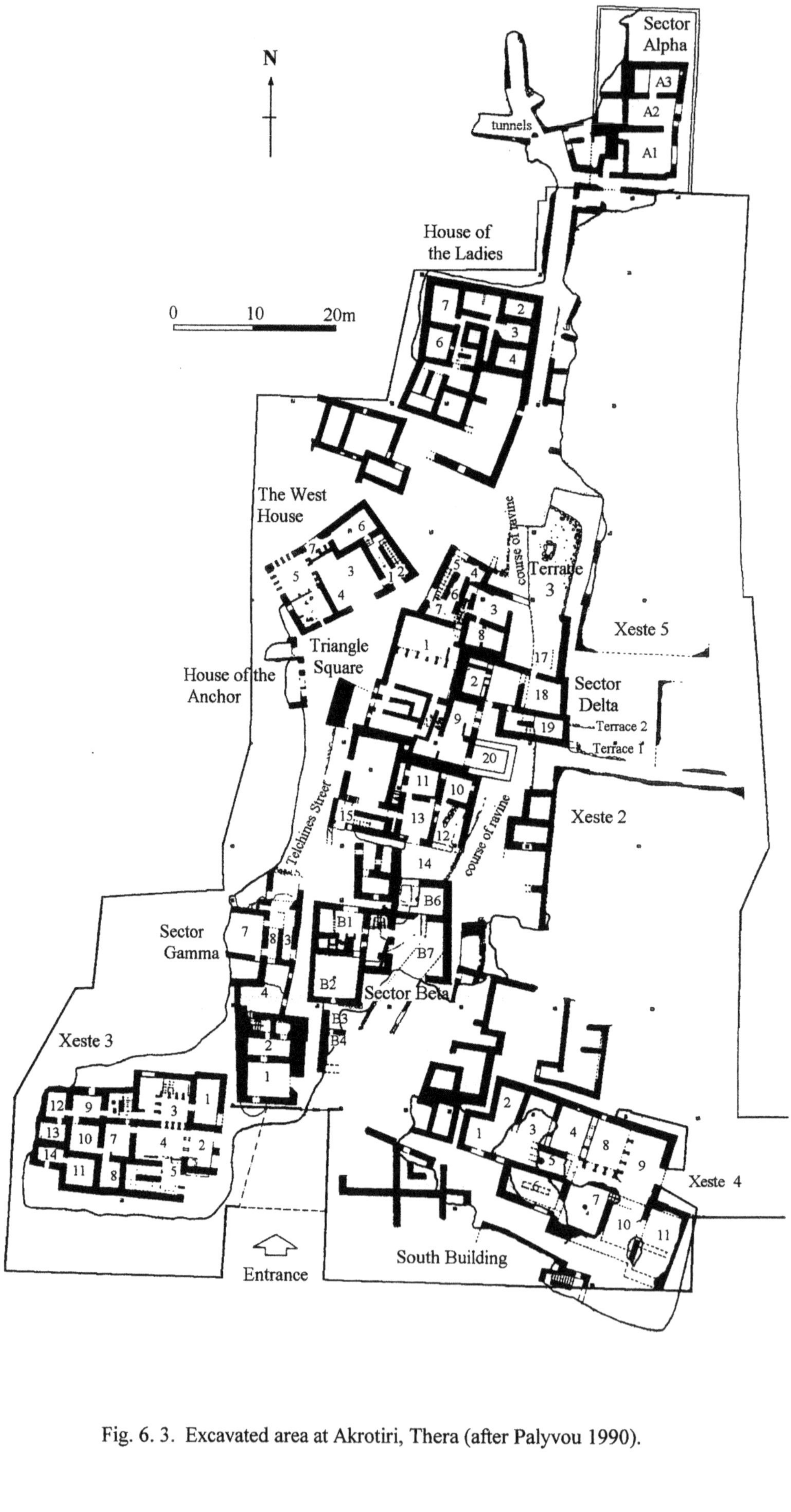

Fig. 6. 3. Excavated area at Akrotiri, Thera (after Palyvou 1990).

Area Sampled in m²	Built Area [a]	Open Spaces	Multiply x 2 If 2 storey
1725 [b]	813	912	1626
10. 000	4713	5287	9426
200.000 [1]	94.260	105. 743	188.520
1.720.000 [2]	810.643	909.357	1.621.286
2.740.000 [3]	1.291.373	1.448.627	2.58.746

(a) built area where wall widths are included in these estimates.
(b) calculations are approximate to $1m^2$ and accuracy decreases for larger denominations, sampling carried out by Sarpaki (1987).
(1) estimate by Ch. Doumas for the extent of the site
(2) catchment area A of assumed extent of site
(3) catchment area A+B of assumed extent of site

Table 6-2 : Estimate of inhabited area in square metres, after Sarpaki (1987).

According to Narroll's (1962) model ($10m^2$/ person), the population of the settlement could have been from a minimum of 9,426 people, if only the first storeys were occupied, to 18,852 people, if both ground and first storeys were occupied. According to the $16m^2$/person model, the population of the settlement has been calculated at 5,891 (1 storey) or 11,782 people (two storeys) (Sarpaki 1987). Sumner's estimates justify 2000-4000 people inhabiting Akrotiri, without taking account of the public places in the settlement, although the figure seems quite low, as the population of Thera before World War II was about 28,000 people; a rough figure of about 10,000 for Late Bronze Age Akrotiri might be a realistic estimate (Table 6.4).

Narrol's Model $10m^2$/person 1 storey/ 2 storey	Model $16m^2$/person 1 storey/ 2 storey	Summer's estimate 100-200 people per settlement ha (2) 1 storey/ 2 storey
81 162	--- ---	---
471 942	294 589	100-200
9426 18852	5891 11782	2000-4000
81064 162128	50665 101330	17200-34400
129137 258274	80711 161442	27400-54800

Table 6-3 : Population estimate based on respective (Table 6.2) inhabited area, after Sarpaki (1987).

The Late Bronze Age destruction

It is suggested from the archaeological record, that there were different phases during the final destruction. Firstly an earthquake took place leading to the first abandonment of the excavated part of the site. It is unknown for how long the area was deserted, but afterwards, some or all of the inhabitants returned to the site, only to leave when further tectonic movements warned them of new destruction, the final one. The violently explosive Plinian eruption, a type named from Pliny the Younger's description of the AD 79 eruption of Vesuvius, destroyed much of the island and covered the rest of it in pumice.

Evidence of clearing earthquake debris from the streets, rebuilding fallen walls as well as restoring wall paintings was found in the settlement. Two clay buckets of fresh mortar for wall-plaster were found in the upper storey, Room 4 of the West House, showing that restorations were in process and suddenly interrupted. Furthermore, Doumas (1983) has suggested that jars with grain were placed under the pier πολύθυρον Room Delta 1 for protection from the earthquake. Marinatos (1939) refers to "troglodytes" that briefly occupied the settlement. Whether it was troglodytes, the owners of the houses, or their servants, and for how long they stayed in the settlement before the final abandonment, is uncertain. The houses were evacuated before the Plinian eruption, since no human skeletons or jewellery have been found at Akrotiri. However, the limited amount of the excavated area does not allow such assumptions. The people could have fled to the countryside, to a nearby island or they could be still found at the bottom of the sea, or somewhere nearby the settlement. During the most recent volcanic eruption in 1926, the inhabitants of Akrotiri found shelters in rock cavities around the mountain of Profitis Ilias. Some of them died of suffocation when pumice from the volcano closed the openings of the caves. It was thought that during the eruption of Vesuvius in AD 79, most of the inhabitants of Herculaneum had escaped (Jashemski 1979, 611), but they appear to have fled to the beach, where many bodies have recently been found beneath the ash flow (cf. Capasso & Ditota 1998).

The town of Akrotiri

The plan of the town of Akrotiri is very much like one of the Cycladic villages today. Narrow streets have been, since prehistory, a deliberate attempt to prevent winds and intruders. The streets were paved with sizeable stones, and underneath them ran the sewers of the settlement. They are connected to pits beneath the streets that receive effluent from the lavatories of the houses through clay pipes built into house walls. The best example of a lavatory was preserved in the West House, at the south-west corner of the first floor (Doumas 1983).

The basic characteristic of the houses of the settlement was a window next to the doorway (παράθυρο; παρά την θύρα). Downstairs, where foodstuffs were stored, the windows were small, while on the upper storeys, which functioned as living quarters and workshops, large windows let light and fresh air in. The material of which the houses are made, is largely unworked stone, and clay and timber used to strengthen the walls (Xeste 2, Xeste 3). Ashlar masonry was used as a frame for doors and windows, for external friezes or to face special buildings (Xeste 4). Clay with straw was used for internal partitions and upper storeys. Wood was used for doors and windows, while the staircases were made mostly of stone and, on rare occasions, of wood (Doumas 1983). The ground floors were of beaten earth, or sometimes of broken sea-shells. The vestibule of the house was made of slabs (e.g. Xeste 3, Mill House of Sector Alpha). In Room Alpha 2 (Pithoi Storeroom), the floor was made of beaten earth mixed with crushed *Murex* shells. In contrast, *Murex* shells were very rarely used for the upper floors. Either a layer of earth on the top of branches and reeds was utilised as a floor or, occasionally, flagstones were arranged on the top of the earth. Very rarely a kind of mosaic made of sea pebbles and a type of cement was also used for the floors (Room Delta 8 and Delta 18). The outside and inner walls of the basements were covered with a mixture of clay and broken straw. In the upper storeys, a kind of plaster was used, that was smoothed while wet with the use of sea pebbles (Doumas 1983).

The ground floors were used for storerooms or workshops. Large jars, pithoi, were used as food containers. Sometimes they were placed against the wall, supported or separated by vertical stone partitions; at other times they were partly buried in the ground (Pithoi Storeroom) or even incorporated in built benches (Room Beta 1). Tools such as mortars, grindstones, hammers, etc., found in the houses suggest that each house had a small workshop that served the everyday needs of its residents.

The West House and other Buildings

The West House (Fig. 6.4) is currently the only fully excavated free-standing building. Two storeys are preserved (Fig. 6.5), and there may have been a third. The ground floor had seven rooms, each room having a door that leads to the next one. Circulation around the house can be achieved both clockwise and anticlockwise. The use of the room 3B is a mystery. Room 3A had a milling installation and Rooms 3C, 5, and 6, according to the archaeological remains, were used for storage. In area 7 was situated a small staircase that lead to the upper floors. Rooms 5 and 6 were used exclusively for storage while rooms 3A, 3C and 4 may have functioned as workrooms. At the first floor, access to the rooms could be gained only clockwise. Room 3 was the largest one, probably dedicated to weaving, in view of the

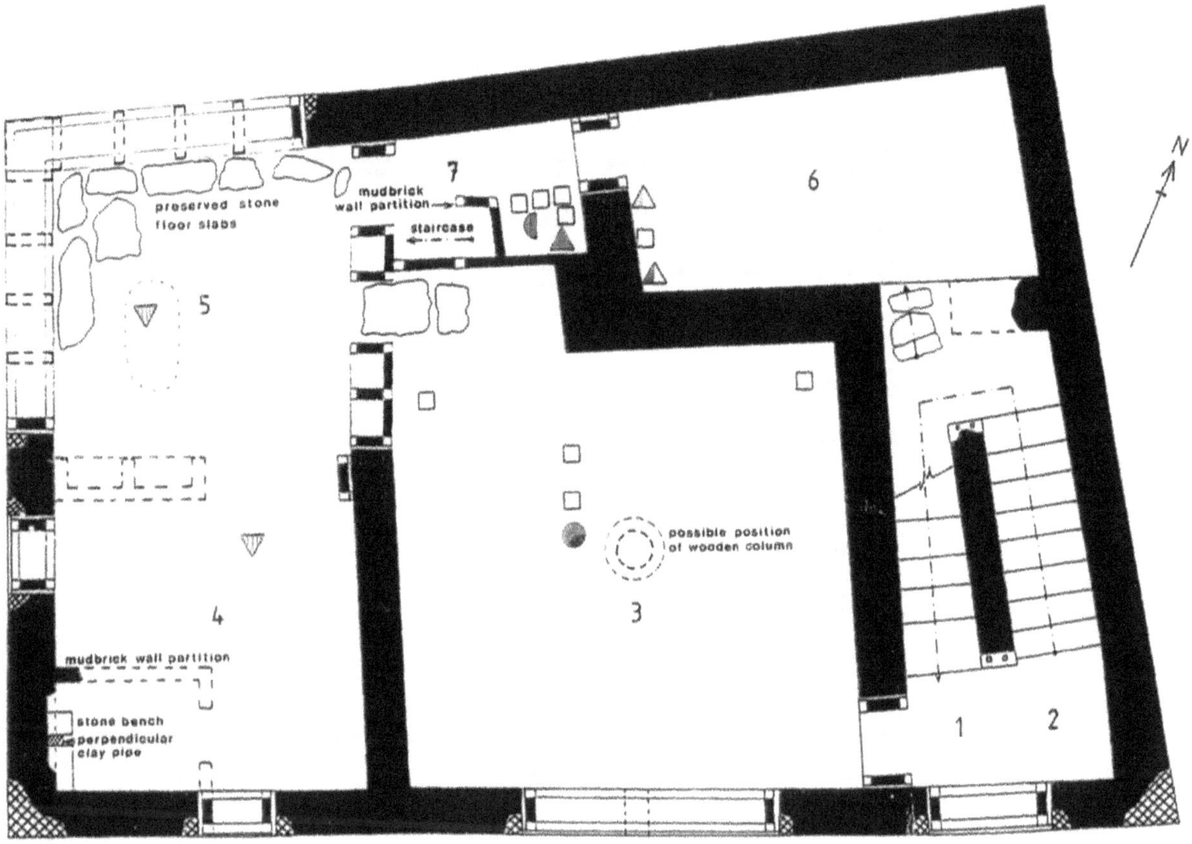

FIRST FLOOR PLAN

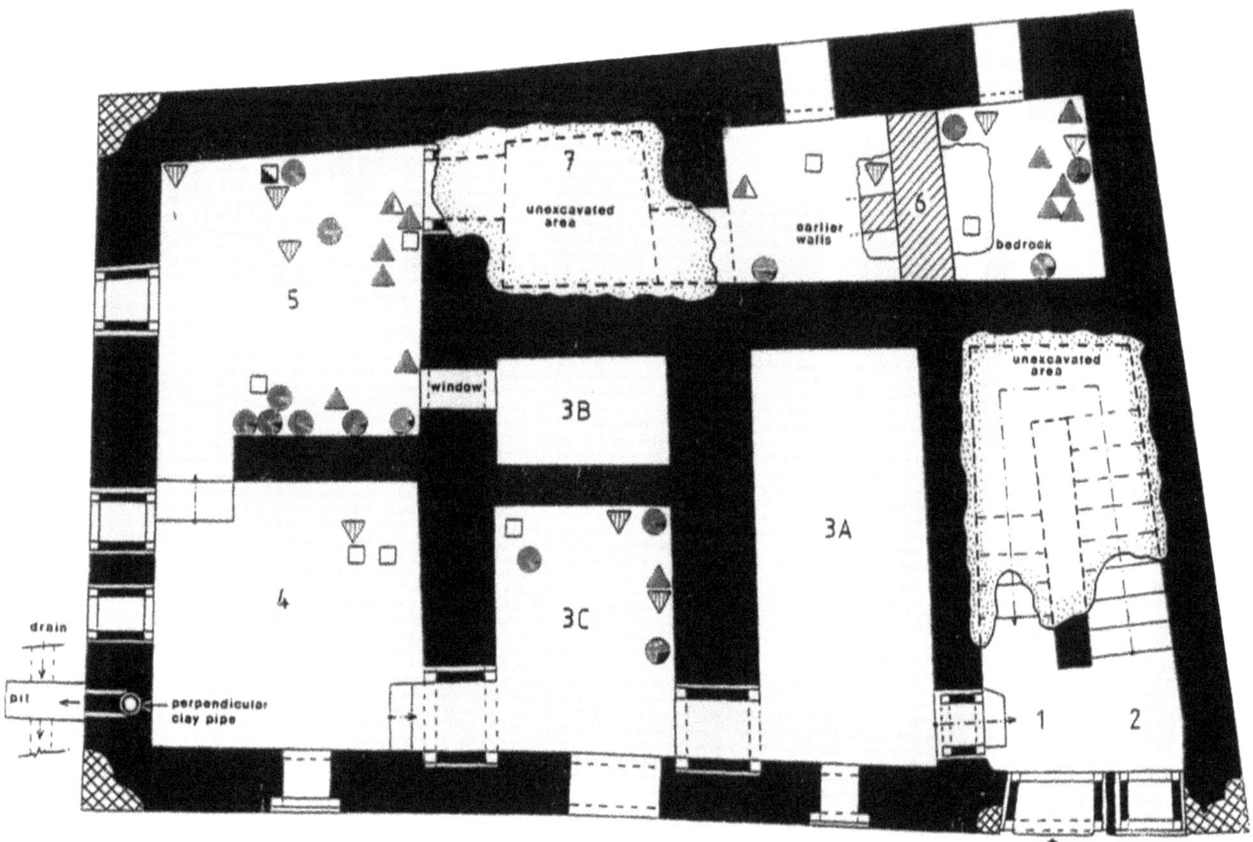

GROUND FLOOR PLAN

Figs. 6. 4 & 6. 5 : Ground (lower) and first (upper) floor plans, West House, Akrotiri, Thera.

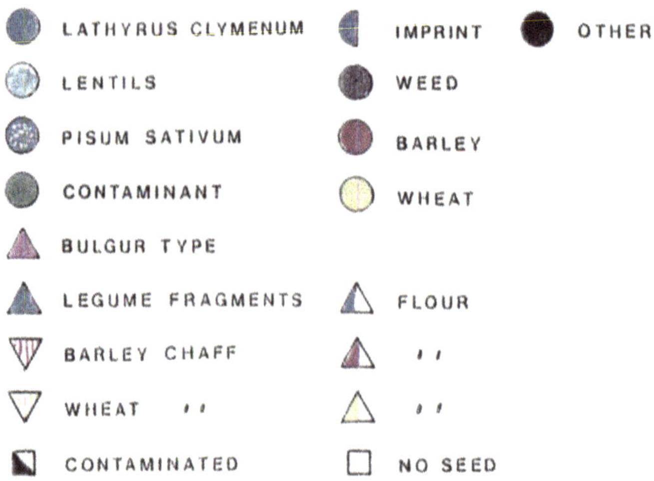

Key to figs. 6. 4 and 6. 5 : Distribution of plant remains within the West House, Akrotiri (after Sarpaki 1987).

Fig. 6. 6: The West House, Akrotiri, Thera.

large window on the south wall, and the loom weights found there. Room 5, next to Room 3, was the most luxurious one of the house, with wall paintings of the two fishermen, the flotilla and the town fresco and the πολυπαράθυρο, pier and window partition, overlooking the west and the north (Sarpaki 1987). Uncertain also is the use of Room 4, while some kind of storage was taking place in Room 6, according to the botanical evidence (*idem*). Area 7 was occupied by the staircase, a small area probably used for storage, and a cupboard, as well as a corridor leading to Room 6. A terrace may have existed above the first floor of the West House. However, only the walls of Room 6 were thicker than other rooms of the first floor and the principal staircase continued in wood to the terrace. So a second floor may have existed just over Room 6.

Two trenches have been dug in front of the West House. The first one along the South wall, 0.86 m from the Southwest corner of the building. The trench was made to test the view that the West House was constructed during the Middle Cycladic period, and that it was repaired at the end of that period and, as a result, the level of the triangular square in front of the House was raised. Three architectural phases were revealed: an Early Cycladic one (wall b), an Middle Cycladic I one (wall a) and an Late Cycladic I/Late Minoan I phase (the West House itself). The second trench was opened along the east wall of the House, 4.06m from the north-east corner of the building. At a depth of 1.80m., a wall was uncovered which was probably a part of the Middle Cycladic I wall (a) of the previous trench. The pottery collected at that level is contemporary with the settlement.

Fig. 6. 7. Delta Building, Akrotiri, Thera.

Excavations at the House of the Ladies have been concluded recently and rooms 2, 3, 4, 6, 7 and 8 have been completely uncovered. The excavation of Room 7 indicates that, at least in this area, the house had three storeys. Wooden beams and reeds were used in construction as revealed by imprints. The wall paintings of the ladies and the daffodils were found in the house. Amongst the finds were large quantities of pottery and stone vases.

Next to building Beta is the enormous building Delta, part of which is still under excavation. It seems that the building is the product of at least five additions to the original structure. At ground level there are at least twenty rooms. In Room Delta 2 was found the wall painting of the lilies (Fig. 6.6), Room Delta 9 was full of pottery, and Room Delta 16 may have been a shop for pottery and other commodities.

Xeste 3, has been fully excavated. The building has at least fourteen rooms at ground level. Room 14 was the last one to be excavated. Small pieces of wall paintings, a sealstone made of black volcanic rock, pieces of pottery, bones, molluscs, charcoal, flint pieces were recovered. The building could have been a public building as the pier windows, the wall paintings and the lustral basin suggest. There are two very well preserved storeys and the staircase in the north wing suggests the existence of a third.

Xeste 4 is a three floor building currently under excavation (PAE 1992). The whole building was faced with ashlar masonry. Its large dimensions, impressive exterior and its numerous wall paintings are again suggestive of use as a public building (Doumas 1992, 176).

Cultivation past and present

According to Doumas (1983, 129), the pre-eruption landscape around Akrotiri was flat and well protected from the north winds by the volcanic cones; it was therefore very suitable for agriculture. The slopes of Mt. Loumavari and Mt. Profitis Ilias could have provided grazing for livestock. Agricultural production in Akrotiri could have ensured self sufficiency, but we should expect that the population would have relied heavily upon fishing. The fishbones found in Akrotiri are of very small fish, which indicates that the marine resources did not contribute in any great extent to the diet of the city, although off-site processing can introduce a significant bias into such assemblages (*cf.* Barrett 1996).

Until recently on the island, the diet was largely based upon barley, pulses, olives, oil, grapes and other fruits and nuts. The consumption of meat was rare, consumed once a week, on Sunday in the Cyclades and most parts of Greece. This pattern of dietary habits was typical of all the Cyclades; Wagstaff *et al.* (1982), however, appear to overestimate the consumption of wheat in the islands.

Archaeoenvironmental studies

Archaeozoology

The main animal species represented in the Akrotiri faunal assemblages are sheep, goat, pig and cow. Also dog, red deer, small equids, hare, fish and birds were present, but in smaller numbers (Bovidae; 8,5%, Capridae; 69.6%, Suidae; 18.3%, Canidae; 0.5%, Equidae; 0.03%, Wild animals 2.8%) (Gamble 1978). Only two human bones have been recovered from the site. However, it should be stressed that in these data smaller animals and fish are under-represented and, since the use of water sieving for biological remains has been utilised, a different picture of subsistence in the settlement has emerged. For example, while previously from 7,155 bones only 35 belonged to fish species, after the use of water sieving in samples from the West House, of 10,607 bones, 1,687 were of fish (Trantalidou 1990). Also the preservation of fish and bird bones is far worse than the preservation of large animal bones. Recent research on the fish bones has been undertaken by Mylona, but the results are not yet available. The animal bone material studied is very fragmented and much of it had been incorporated in structures. Therefore a study based on MNI (minimum number of individuals), and not on NISP (number of identifiable specimens), should give a more realistic picture of the dietary habits and the economy of the settlement.

The study of animal bones from the site has shown that cattle were not bred intensively, while there was a large number of sheep. Cattle in the Cyclades at the beginning of this century, were used mainly for traction. As the soils of Thera are light, asses or horses were suitable for the same purpose, but their small numbers are not indicative of such an activity. Gamble (1978, 746) suggests that probably the settlement relied on dairy products, as well as the meat of surplus lambs. Some sort of pasture management should also be assumed, perhaps incorporating olive or other fruit trees. In conclusion, on the limited present evidence, sheep and goats were mainly kept for their milk and most probably wool, while cattle were possibly used for traction.

Molluscs

A large variety of mollusc species has been recovered from Akrotiri. 54.1% of the assemblage studied was composed of shells of *Murex trunculus* and *M. brandaris*. The number of shells found (2,153 marine, 196 terrestrial), according to Karali-Yiannakopoulou (1990), is quite small for such an extensive site. The molluscs were mainly recovered from floors and walls and were very fragmented. Unfortunately, their study was not based on the minimum number of individuals, and work has concentrated on the larger shells. In addition, the environ-mental reconstruction was limited as the molluscs were distinguished into two main groups only, marine and terrestrial.

Many of the molluscs have been found outside buildings, in the streets with debris and only few in the rooms. Certain molluscan species were probably used as bait, for example, the limpets, *Patella* sp., or were eaten as human food, such as the large land snails, *Helix* sp. The very abundant *Murex* shells were mostly crushed. *Murex* is used for food, but is also the source of the famous purple dye, Tyrian purple, πορφύρα. The dye is found in the hypobranchial gland of the mollusc in the mantle cavity, and the extracted liquid is whitish in the beginning, and, after being in contact with the air and sun, becomes purple. For the extraction of the dye, large numbers of the molluscs are required, as well as basins for boiling, and installations to dry them in the open air (Robinson 1969). Such an activity may have taken place somewhere outside the settlement. The role of molluscs in the diet and economic life of Akrotiri is still under study.

Animal representations in Theran art

The animal world is depicted in Theran art. In the Akrotiri wall paintings, there are representations of swallows, monkeys, caprines, dolphins, deer, antelopes, lions and other animals. The detailed character and the accuracy of the work in most is often surprising. Most of them are still part of the fauna of the island, for example, the sheep and goats. Dolphins are also typical animals in the Mediterranean, linked with sea and maritime activities. Migrating swallows depicted in paintings, even today are an indication of springtime or the sailing season.

Fallow deer (*Dama dama*) and fish (the fisherman on the west wall of the West House, Room 5, carries a mackerel, *Coryphaena hippurus*), ducks and other birds, moths and dragonflies are some of the animals depicted in the wall paintings. As some of the animals still exist on Thera, we can assume that the artist was depicting animals with which he was familiar. The so-called butterflies and the dragonflies depicted in the wall paintings will be discussed in detail elsewhere.

However other animals not typical of the Aegean are also represented in the Theran wall paintings, for example, the monkeys. The discovery of a Bronze Age skull of *Cercopithecus* sp. in Thera (Poulianos 1972, 229-230) could mean that the depiction of monkeys in the Theran wall paintings is the result of observation of the real animal. However, the identification of the find has been questioned. It has been claimed that the monkey had arrived in Thera as gift, like those represented in the expedition to Punt at Deir el-Bahri and at the tomb of Reckhmire in Egypt (McDermott 1938, 3-6). Contrary to what Van Schoonwinkel (1990) believes, the animal was not a part of the local Egyptian fauna (Van Schoonwinkel 1990, 336), and probably originated in Nubia (Reynolds 1968). The cheetah, or serval cat, is another animal depicted in the "Nilotic landscape" that does not belong to the Theran fauna. Evans (1930, 113-118) and Immerwahr (1985, 41-50) argue for an exotic scene as the palm trees, the papyrus, and the felines are part of the Egyptian scenery. In Egypt, from the Vth Dynasty, the mastaba of Ti and the tomb of Nianchchnum and Chnumhotep at Saqqarah have representations of felines of several types (Van Schoonwinkel 1990, 338). Morgan (1987, 42) believes that the animals are wild cats that occurred in Crete, *Felis silvestris (=agrius)*. Antelopes are also exotic animals in the Aegean. Their representation in the Theran wall paintings could be a theme borrowed from the Egyptian iconography (Stork 1975, 319-323; Brunner-Traut 1977, 426-427), although the quality of the illustrations suggests some familiarity with the animals. Morgan's (1987) opinion that the miniature wall paintings mainly depicted the Aegean fauna is over-optimistic from the past distribution of the fauna point of view.

In conclusion, in the Theran wall paintings are represented animals of importance for the life of the Bronze Age settlement. Some of them are parts of the fauna of the island connected with everyday scenes and activities, while others show strong relations with Syria, Anatolia and Egypt.

Deforestation

According to Rackham (1982, 177), if Greece was left untouched by man, the climax vegetation would have been forest, evergreen and sclerophilous in low rainfall areas with *Q. coccifera*, *Pinus* sp. and *Ceratonia siliqua*, and in higher altitudes *Quercus*, *Pinus* and *Abies*. During the Bronze Age, there was a massive human impact on vegetation for both arable and pastoral cultivation. The industries which consumed large amount of wood, pottery, metalworking, lime-making, as well as the needs for timber for architecture and shipbuilding were critical for the forests (Wertime 1983, 452). The amount of timber for ships, as illustrated at the Akrotiri miniature fresco for example, must have been considerable (Meiggs 1982, 97). Extensive woodworking is recorded in the Linear B tablets from Pylos, where the guilds of ship-builders and carpenters as well as wood-cutters are mentioned (Ventris & Chadwick 1973, 123).

Human impact on the environment and deforestation are connected with settlement patterns and population numbers. Although no suitable deposits for palynological study have been found, it is apparent that the process of deforestation had started long before the destructive eruption on the island of Thera, and the demands of the settlement would have meant significant changes in vegetation patterns.

Charcoal

From the analysis of charcoal at Akrotiri have been identified the following woods: olive (sample P-1601), pine (several samples) (Weinstein & Betancourt 1978, 810), and tamarisk (Sarpaki 1987). However no systematic work on charcoal fragments has yet been undertaken, and there is therefore no way of knowing the types of wood used in the settlement in detail, or the ways different wood species were used. *Arundo donax,* the giant reed, was extensively used in architecture, and many impressions of the reed have been recovered. Systematic research on charcoal fragments from the settlement would help to obtain a clearer picture of the patterns of use of wood.

Evidence for trade

Weights and measures

Akrotiri has revealed the largest collection of prehistoric weights found in a single settlement in the Aegean, consisting of 102 lead pieces and three stone ones (Michailidou 1990, 408). Petruso (1979) believes that the weight system at Akrotiri corresponded to the Minoan one. However, as some of the lead weights do not seem to fit the Minoan system, he assumes that two systems were in use at the same time. Further research by Michailidou (1990) has shown that there is no substantial evidence to support the existence of a separate Cycladic metrical system. On the contrary, there appears to have been one metrical system for trade, the one recorded in the Linear B tablets (Michailidou 1990, 417-418), and it is argued that lead weights were not only "tools of trade" (Petruso 1979, 139) but objects of trade in themselves (Michailidou 1990, 417).

The commodities exchanged could have been cloth, food crops, dyes, wax, medicines, oils, and perfumes. For example, it is possible that *Crocus cartwrightianus* or *Crocus sativus* was exported as well as linen, wool or even silk, and condiments such as cumin, coriander, *Cyperus*, sage rose, fennel, celery, cumin, cardamon, sesame, safflower, and mint. Wool and textiles were measured in 'wool units' (LANA = 3kg). The existence of small weights of 25g (Michailidaou 1990) is also of interest. Perhaps, very precious, light commodities were measured with such weights, such as fine linen, silk or cotton thread and saffron. The coexistence in the same house (for example the West House, Sector A) of a collection of lead weights and a large quantity of loom weights justifies the belief in a specialised textile industry at Akrotiri.

The fact that the same metrical system was used in the Cyclades and Crete (Petruso 1979, 138; Michailidou 1990), and the existence of other metrical systems at the same time, suggests different kinds of trade or exchange (Parise 1986). As organic materials are rarely preserved archaeologically, the abundance of weights at Akrotiri is the strongest argument for activities connected to trade in the island during the Late Bronze Age.

Loom weights

Over 950 loom weights have been recovered so far from Akrotiri, and more than 450 of them were found inside the West House (Tzachili 1990, 381). The presence of such a large number of loom weights in certain rooms of the West house, Sector A and buildings B and D argues for specialisation on the site in textile manufacture (Tzachili 1990, 385). Among the loom weights, there are a number of very light ones (*idem*, 383) that were probably used for the weaving of the fine textiles represented in the wall paintings. For example, the theme of the eastern section of Room 1, at the House of the Ladies has been interpreted as a robing ceremony; the female figures wear luxurious often transparent garments, and, even if not a ritual act, the scene clearly demonstrates the importance of luxury clothing items as symbols of prestige in the life of the settlement.

Conclusion

The excavation of the settlement of Akrotiri, destroyed by a Plinian eruption, during the middle of the second millennium BC, has revealed a wealthy, flourishing community. In the part so far excavated, no palatial building have been identified. Several buildings, private and public, have been uncovered. The archaeological finds are impressive, and show the special role of Thera in the Cyclades and the Aegean, and relations, probably through trade, with Crete, Egypt and the Levant.

Years BC	Egypt Dynasties	Crete	Mainland Greece	North - East Aegean	Cyclades	Thera
	XIX	Late Minoan III 2	Late Helladic III B	Troy VII	Middle Helladic III B	Late Cycladic III B
1300		III 1	III A		III A	
1400	Amarna	II ?	II		II	Gap
1500	XVIII	I B		Troy	I B	Eruption
					I A	Late Cycladic III A
1600	Hyksos	I A	I	VI		
1700	XIII - XIV	Middle Minoan III A			Middle	Middle
		II B				
1800	XII	II A			Helladic	Cycladic
1900		I B				
2000	XI	I A		Troy V		
2100	First Intermediate Period	Early Minoan III	Early III	Troy IV	III	Early Cycladic III
2200		II		Troy III		
2300						

Table 6-4. The Bronze Age in Egypt and the Aegean, outline chronological table (modified from Doumas 1983 & Cline 1994).

7. Akrotiri-The Archaeoentomology

Introduction

Akrotiri provides excellent opportunities for archaeoentomological work. Not only is it one of the most extensively excavated site in the Cyclades, but also the Plinian eruption that took place at the beginning of the Late Bronze Age covered everything with a thick blanket of tephra, ensuring that the preservation on site is excellent. Previous research upon fossil insects from Santorini is restricted to Freidrich's (1980) finds of the olive whitefly, *Aleurolobus olivinus* Silv. on the leaves of olive preserved in a palaeosol dated to about 37,000 BP. The systematic archaeoentomological study under-taken here, is a pioneer project in the Mediterranean region and shows the potential of insect research on archaeological sites in the region.

Objectives

Various questions were raised at the beginning of the archaeoentomological research. In summary, the aims of the insect study from Akrotiri are as follows:

a. Recovery of insect fossil records.
b. Recording the species of insect pests in the storerooms and other contexts of the settlement and comparison with the modern fauna of the island.
c. Recording infestation levels in pulses and discussion of their economic implications
d. Research into the possibility of the use of insecticides in the storerooms
e. Reconstruction of environmental conditions from the insect fauna.
f. Reconstruction of human activities on the site based on the archaeo-entomological finds.

Choice of area studied and sampling methods

As many contexts as possible were sampled for insects, a total of 267 samples. However, due to the amount of time required to process samples, and as the West House was almost fully excavated when the study was undertaken, the largest amount of the work was dedicated to the study of material from this particular house. The archaeobotanical study from the House had already provided a list of plant species found mainly in the storage areas, and noted insect infestation in the pulses (Sarpaki 1987). A further, more detailed study of the material was therefore considered necessary.

Samples were taken from as many contexts as possible. In the beginning of the study, all the samples taken were sorted under the microscope. However, as the amount of floor samples and sediments from other contexts was huge, and even after dry sieving, the quantities to be sorted remained unrealistic in the period of time available, the study had to be restricted to the contents of pots and samples, where concentrations of preserved botanical material indicated the probability of insect remains.

Concentrated botanical material was sorted in full, while a minimum of three litres was processed in other cases. For a large part of the samples, the material had been collected during the early years of the excavation, when Sp. Marinatos was the site director (1967-1974). This was utilised, as well as other material already collected and processed by the archaeobotanist at the site, Anaya Sarpaki. From 1989 onwards, samples were taken for insect analysis. However, the presence of the researcher on site was not always possible, and the collection of samples relied up on the good will of the excavation director, C. Doumas.

The West House

The archaeobotanical study of the West House resulted in exciting finds. Plant remains were found both on the ground and the first floor of the building (Fig. 6.5), and in some cases, heavily infested plant remains gave evidence of the activity of insect pests. The samples from the West House produced mainly synanthropic species. This is the first time that insect assemblages have been systematically recorded from a Mediterranean archaeological site. The general characteristics of the synanthropic faunas are discussed later.

Ground Floor

The great majority of the insect pests found, was recorded from stored crops mainly from three rooms: 5, 6 and 3C. Most of the pulses were kept in Room 5, and they were "fava" seeds, *Lathyrus clymenum* and *Lathyrus cicera/sativus*. From the botanical samples studied by A. Sarpaki, eight were of pulses, five were bulgur type, one of legume fragments, and one of chaff (Sarpaki 1987, 219). Lentil crops were kept in Room 6 (sample 37/61), and Spanish vetchling and pea crops in Room 3C (samples 65&31). Room 3A had a milling installation, and most probably Room 3C was an auxiliary to Room 3A, used for processing and perhaps for cooking (Sarpaki 1987, 220-221).

Room 3B has no door, just a window facing into the main storeroom, Room 5. The room does not seem to be a cistern, as it is unlined. However, it is possible that Room 3B was for crop storage. The fact that there is no door would be an extra measure for safe storage against pests. As it is accessible only through the main storeroom from a window, it might well have served for the same purpose. Room 3B brings in mind Egyptian storerooms with small openings, or even near contemporary ethnographic parallels in household storerooms in different cultures, constructed behind walls or under floors, accessible through a small door.

First Floor

Room 3 and Room 5 could have been living and working rooms. Room 5 could also have been a kind of a living room, "an entertaining-cum-religious area". Room 4 was a more private room, perhaps used for sleeping. Room 6 is near the staircase, and is therefore easily accessed from downstairs. This produced three botanical samples, the first two from storage jars which had been built into a bench, and it can be assumed that the room was designed for storage (Sarpaki 1987). Room 7 was the linking part between the ground floor and the first floor, partly used for storage, and included samples identified as bulgur and *Lathyrus clymenum* (*idem*).

The archaeoentomology of the West House

Grain Fauna

The insects from the West House are listed in tables 7.1 and 7.2. In the absence of a Greek checklist, the nomenclature follows that of Lucht (1987). Most of the insects listed are synanthropic, only able to maintain breeding populations in close association with Man.

Taxon				
	Room 3c	**Room 5**	**Room 6**	**Drain**
Coleoptera				
Carabidae				
Amara sp.		1		
Scarabaeidae				
Onthophagus sp.			1	
Tropinota hirta (Poda)		1		
Dermestidae				
Dermestes sp.		1		
Anobiidae				
Stegobium paniceum (L.)			1	
Bostrychidae				
Rhyzopertha dominica (F.)			1	
Silvanidae				
Oryzaephilus sp.		1		
Cryptophagidae				
Cryptophagus sp.				1
indet.		1		
Tenebrionidae				
cf. *Guranus* sp.		1		

indet.		1		
Bruchidae				
Bruchus rufipes Hbst.		323	1	
Curculionidae				
Hypera sp.		2		
cf. *Hypera* sp.		1		
Sitophilus granarius (L.)	2	15		
Troglorhynchus cf. *anophthalmus* Schm.		57	1	
Diptera				
Spharoceridae				m
Phoridae				m
Ephedridae				m
Scatella sp.				m
indet.				m
Hymenoptera				ff
Formicidae	+	+	+	
indet.		+	+	1

ff = fragments m = many

Table 7-1 List of species from the West House, Akrotiri

As expected, the pests in the pulses were Bruchids, *Bruchus rufipes* (Hbst.). Most Bruchids infest seeds in the field. Eggs are laid on the young legume bean pods, and the larvae bore into the developing seeds. The life cycle is completed in the seed and adults emerge by biting their way out. The adults fly readily and they are common on legumes in springtime. Herford (1935) records *B. rufipes* as a pest of common vetch, *Vicia sativa* L., and species of *Lathyrus*, and notes its distribution as "Trans-Caucasian". Harde (1984) comments that it is the most common member of the genus in Central Europe, and it was noted as frequent in modern samples of legume crops from the Greek mainland (D. Smith, pers. comm.). *B. rufipes* is also included in a checklist of the most common pests of crops in Greece prepared by Pelekassis (1962), where it is recorded from the horsebean, *Vicia faba* L., and stored leguminous seeds.

Other pests, typical of storerooms, are also present. *Sitophilus granarius* (L.), the grain weevil and *Oryzaephilus* sp., the saw toothed grain beetle, as well as *Rhyzopertha dominica* (F.), the lesser grain borer, *Stegobium paniceum* (L.) the biscuit beetle, and *Dermestes* sp., a bacon or hide beetle occur in the assemblage from the storerooms. As *S. granarius* is flightless, its occurrence indicates transport to the island, at some point in time; it could even have arrived in the seed corn of the earliest settlers. *Rhyzopertha dominica* (F.) has been thought to be of South American origin (Munro 1966), but this record from the beginning of the Late Bronze Age, indicates an Old World origin; Zacher (1934) also recorded the species from a roughly contemporary pot in Egypt, but failed the specify the location of the find.

Lasioderma serricorne (F.) is another storage pest, here recorded for the first time in a non-Egyptian context. Elytra of Dermestids were also recorded from the West House. The identification is only possible to the generic level. Species of *Dermestes* are mainly carrion feeders often occurring on meat, skins and furs. Galleries in the wood samples also indicate the presence of wood borers, possibly *Anobium* sp. However, the samples were not big enough to distinguish pattern and provide a more certain identification.

The most unexpected find from the West House, was an eyeless troglodytic species, *Troglorhynchus* cf. *anophthalmus* (Schm.), originally recorded under thick leaf litter, and never before found in an archaeological context. Pelekassis (1962) does not list the species as a pest, and members of this genus are largely troglodytic (Jeannell 1926). The taxonomy of the genus is complex, with endemic species claimed from many caves (*cf.* Giordan 1988; Osella 1974; 1986; Osella & Abbrazzi 1985). Freiser (1981) notes two species, *T. anophthalmus* and *T. baldensis* Czwal., from northern Italy, Austria and the former Yugoslavia; both were sieved from deep leaf litter or the humus-rich root layer, although only the one, *T. anophthalmus*, is included by Koch (1992). The presence of sixteen individuals in intimate association with the pulses infested with *Bruchus rufipes* implies that the animal had found a suitable *pabulum* in the stored product. Its habitat is not the dry conditions, which one would normally associate with the preservation of pulses in large jars, and the pattern of storage requires further examination. The weevil perhaps implies storage in liquid. If that was the case, the conditions in the storerooms were not that different from its original *pabula*, and therefore the weevil could have adjusted with no problem. It is possible that *Troglorhynchus* species were attracted simply to poorly dried stored crops, but the blind weevil seems an unlikely denizen of the settlement in general, and is not present in other samples. The possibility exists that it may have been inadvertently transferred with crops from storage elsewhere, either in caves or in pits. As the island is volcanic, rock cavities are a common feature and the inhabitants exploited them for storage purposes well into the present century. It is highly possible that the weevil already dwelt in the cavities used for storing crops, found new food sources, and became a secondary pest thereafter. The possibility of the caves being used for bulk storage should be considered, although the botanical finds argue for storage at the household level. Perhaps communal storage was taking place outside the settlement for security reasons. Transfer of produce from temporary cave or pit stores in the rural hinterland of Akrotiri provides a possible route for *Troglorhynchus* to become at least a temporary stored product pest in sufficiently cool and damp locations. Another explanation is that storage in liquids, such as olive oil or oil lees (*amurca*), or even vinegar, had been practised in Akrotiri. Steeping the pulses in oil, vinegar or *amurca* has been an effective and well known measure against pests from Antiquity to modern times. It is also possible that, like several of the other pests, it arrived in cereals or other damp produce in the holds of Bronze Age shipping. The primary habitat of the genus could have been the deep, damp litter of the forest floor and, as forest clearance has proceeded, populations became increasingly restricted to isolated localities. Without detailed knowledge of the present distribution and habitat requirements of *Troglorhynchus* species, the problem of the Santorini specimens must remain unsolved.

Other species found in the West House are associated with dry environments and *Hypera* sp., a weevil genus associated largely with leguminous plants in the field, is one of the most taxa species in the fields. The dung beetle, *Onthophagus* sp., inside the lentil sample from Room 6, could imply that some kind of processing was taking place outside the storerooms, although another explanation could be that the insect fell from dung that had been used to seal the pot. It is also possible that dung could have also been mixed together with the lentils as an insect repellent. As the sample contained damaged grains, but was free from Bruchid individuals, it is probable that sieving to remove insects had preceded storage.

Samples taken from the toilet of the West House are indicative of foul conditions. Mineralised fly puparia of Sphaeroceridae, Phoridae and Ephedridae, probably *Scatella* sp., (Skidmore pers. comm.), witness accumulations of human faeces. The fact that they were killed abruptly (most of them had not emerged, but were killed during the larva stage) is either the result of the eruption, or of suffocation that was caused by large accumulation of faeces.

Infestation levels

Infestation levels in the samples of the West House are variable (Table 7.4, 7.5). Pulses infested with Bruchids showed infestation levels from 1% to 9%. However, it should be mentioned that the specific sample infested only to 1%, could be the outcome of before storage treatment for insects. If the inhabitants were using sieving to clean the crop from insects, as suspected on botanical grounds (Sarpaki 1987), the infestation levels found in the samples from the West House are quite high. Considering pre-harvest losses also, it can safely be assumed that significant proportions of the crop was destroyed annually by insect attack. Taking into consideration that every so often the crop could be completely destroyed, either because of insect infestation or drought, an assumption of annual losses of the order of 20% is well justified. At the present day, the use of chemical substances as insecticides has reduced considerably the crop loss due to insect infestation (cf., for example, Baeta Naeves 1971). However, until the extensive practice of artificial methods for controlling pests and optimising the preservation of food commodities, pests were a major problem that could lead to famine. In India in 1916, wheat was reported to be infested up to 33%, and Rahmann (1944) reported 73% of losses in wheat in the Punjab (Tyagi & Girish 1975). Hinds (1914) noted 1% losses monthly of wheat in Central Alabama. Two centuries ago, in France, losses caused by infestation by a single pest, *Sitotroga cereallela*, the Angoumois moth, were one of the reasons for the famine that helped trigger the French Revolution (Buckland 1981).

Taxon	MNI
Coleoptera	
Staphylinidae	2
Anotylus sp.	1
Dermestidae	
Lasioderma serricorne (F.)	2
Bostrychidae	
indet.	1
Cryptophagidae	
indet.	1
Bruchidae	
Bruchus sp.	2
Curculionidae	
Sitophilus granarius (L.)	1
Arachnidae	3
Other	
insect coprolites	m
fungi	m

Table 7-2 West House: Insect remains from trenches.

Sample 1 (*Lathyrus clymenum*)	Head	Thorax	Left Elytron	Right Elytron
Bruchus rufipes Hbst.	155	46	298	281
Troglorhynchus anophthalmus Schm.	20	22	46	47

Infestation level: 8-9%

Sample 5/10 (*Lathyrus clymenum*)	Head	Thorax	Left Elytron	Right Elytron
Bruchus rufipes Hbst.	37	24	21	31
Sitophilus granarius (L.)	2	5	9	8
Troglorhynchus anophthalmus Schm.	2	5	9	10
Sample 16 (*Lathyrus clymenum*)	Head	Thorax	Left Elytron	Right Elytron
Bruchus rufipes Hbst.	8	2	3	3
Sitophilus granarius (L.)		1	5	5

Sample 20/29 (*Lathyrus clymenum*)	Head	Thorax	Left Elytron	Right Elytron
Rhyzopertha dominica (F.)	1	1	1	1
Bruchus rufipes Hbst.		4	1	2

Infestation level: 8%

Sample 28 (broken pulses)	Head	Thorax	Left Elytron	Right Elytron
Rhyzopertha dominica (F.)	1	1	1	1
Bruchus rufipes Hbst.			1	
Troglorhynchus anophthalmus Schm.		1		

Table 7-3 Room 5, West House: Species list and infestation levels.

Another point should be raised concerning the insect pests from the West House. In modern Western cultures, insects are usually considered as something unclean. However, other cultures have a completely different approach towards insects, and consumption of insects is a common way to supplement everyday diet (Sutton 1996). A scattering of livestock in the food creates no problems (*cf.* Osborne 1983), but too many insects can make food toxic (*cf.* Buckland 1982).

Taxon	Head	Thorax	Left Elytron	Right Elytron
Rhyzopertha dominica (F.)	2	2	2	2
Oryzaephilus sp.		-	2	1
Bruchus rufipes Hbst.	200	76	323	317
Sitophilus granarius (L.)	2	6	15	13
Troglorhynchus anophthalmus Schm.	22	25	55	57

Table 7-4 Species list from Room 5, West House, Akrotiri.

Use of insecticides

It seems that in Late Bronze Age Akrotiri, several measures were taken against pest losses. Firstly the environment inside the rooms dedicated to storage was controlled; smaller windows, thicker walls, and walls and floors covered with lime are ideal for good preservation as they provide cool and stable conditions. In some cases, pithoi were built inside benches (Room 6 first floor, Room Delta 1 (Fig. 7.1)), construction that could help maintain a cool environment for the stored commodities. Also airtight storage was used, as there are clear indications of the sealing of the ceramic containers (pithoi and other vessels) with unbaked clay and stone covers, sometimes even stones that fit the mouths of the vessels. The technique, however, relies on the build up of carbon dioxide from the respiring seeds to levels sufficient to suffocate the pests, and, as the clay jars had a porous texture, without coating, they alone would be insufficient to maintain this. The occurrence of the dung beetle *Onthophagus* sp. in the lentils from the West House, has been related to the use of dung for sealing the containers. Dung-lined baskets are known from modern rural Greece (Halstead 1987), and dung mixed with the crop was a method practised at least since 1600 BC against pests (Papyrus Ebers, Ebell 1937). Cow dung was also found together with grain in Tut'ankhamun's tomb (Vartavan 1990).

A variety of other substances were also used as insecticides since very early times, and the first relevant reference is found in the Papyrus of Ebers (*idem*). The possibility of the use of such substances has been examined in the case of Akrotiri (Table 7.8; Panagiotakopulu *et al.* 1995). Several of the usual substances, like sand and ash, could not be proven to be added as insecticides when found together with botanical material, as the taphonomy does not allow us to be certain of the reasons for their presence. However, in other cases, the use of substances as insecticides gives a plausible explanation for their existence in the context. In the case of *Thymelaea hirsuta*, a kind of laurel, the consistent presence of leaves in all the samples from the West House could imply deliberate addition of the leaves of the plant in the stored crops. The plant grows in maritime sandy and rocky places around the Mediterranean (Polunin & Huxley 1965), and it is an unlikely crop weed; it is very common on the island today in the garrigue zone. Both Theophrastus (HP VI, 2.2) and Pliny (HN XIII, 114) refer to the use of the plant as a remedy against the poison of snakes, but it appears likely that it was also deemed to have had insecticidal properties. Curiously, a connection with snakes is borne out by the seeds of another "contaminant", *Coriandrum sativum*, coriander, in one of the samples of bulgur from the West House. These were of such a size that they would have been trapped by the sieving which had evidently been used to remove impurities from the foodstuff, and it is probable that they were deliberately added after preparation. Pliny (HN XXV, 104) describes the plant as an antidote to snake venom. Evidence of sieving is also present in samples of *Lathyrus clymenum* from Room 5. Two samples (16 and 20/29) appear to represent the primary usable grain, and the tailings from the same crop: both had similar weed flora and evidence of Bruchid infestation, the former at less than 1% and the latter at 8%. Amongst the prime pulses were seeds of coriander. All were broken in half, and of such a diameter that they would have passed through the mesh, which appears to have been employed in separating the crop before storage (Sarpaki 1990).

Another feature related to the infested samples was the recovery of insect coprolites. The higher the amount of infestation, the larger the amount of insect coprolites recovered. So, eventually even if not completely devoured, the build up of waste products, including uric acid, would have made the grain unfit for consumption.

Insect fossils from other Contexts

House of the Ladies

Entomological work on samples from the House of the Ladies resulted in a completely different insect assemblage from those from the West House. Very few individuals were preserved (Table 7.5). *Gnathoncus* sp., a predator on fly maggots, has often been found in bird nests as well as stored products (Halstead 1963). Other insect species relate to vegetation. The weevils *Hypera* sp. and *Sitona* sp. are frequent on leguminous plants and the ladybird *Coccinella* sp., feeds on aphids. *Aphodius* spp. are largely found in dung and rotting plant debris. *Ctesias* sp. - only one species *C. serra* (F.) occurs in Central Europe (Koch 1989) - are carnivorous, and are often found in bird's nests or under loose dry bark of broad-leaved trees such as oak, elm, ash, willow. Adults and larvae can also be found amongst spider's webs where they feed on insect remains and detritus (Hyman 1992), and the species may have found suitable habitats in the house. *Bruchus* sp. is a pest on pulses. The fact that only one was found in the House of the Ladies is either a result of insufficient preservation, or that limited storage was taking place in the house. *Tribolium* sp., recovered also from Room 2, however, testifies to the existence of cereal products in the house. The insect is a secondary pest of stored products and is typically found in flour (Munro 1966). The presence of insect faeces in samples from the House of the Ladies, suggests some impact of insect pests during the storage of food commodities.

Taxon	MNI
Histeridae	
Gnathoncus sp.	1
Scarabaeidae	
Aphodius sp.	1
Dermestidae	
indet.	
Ctesias sp.	1
Bostrychidae	
indet.	1
Coccinellidae	
indet.	1
Tenebrionidae	
Tribolium sp.	1
Bruchidae	
Bruchus sp.	1
Curculionidae	
Sitona sp.	1
Hypera sp.	1
Diptera	
puparium	1
Arachnidae	1

Table 7-5 Insect fossils from the House of the Ladies, Akrotiri.

Xeste 3

Only a few samples have been processed from this structure., an apparently public building. These have produced only a single calcified fly puparium, and preservation overall is poor.

Taxon	MNI
Diptera	
puparium	1

Pessos 17

Pessos 17 is a structure of unknown use, although the insect fauna from it suggests that it is an open space environment (Table 7.7). The species list from Pessos 17 consists of open air species as *Enoplorum* sp., *Oedescelus* sp., *Brachyceres* sp., *Cetonia* sp., and elytra of large Tenebrionids. The chafers, *Cetonia* sp. have adults which may be found on flowers, fruit trees; sometimes they may become pests on fruit trees. (Horion 1957). The dung beetle, *Aphodius* sp., was also present.

Taxon	MNI
Coleoptera	
Scarabaeidae	
Aphodius sp.	1
Cetonia sp.	1
Tenebrionidae	
Tenebrio sp.	2
Enoplorum sp.	2
Oedesculus sp.	1
Curculionidae	
Brachyceres sp.	4

Table 7-1 Insect fossils from Pessos 17, Akrotiri.

Delta Building

Excavation has continued in building Delta. Room 1, the storeroom of the structure, produced large pithoi containing botanical material, and the material was identified on site by the archaeobotanist of the excavation, A. Sarpaki, as flour. Samples sorted immediately afterwards had shown no insect infestation at all, but in the laboratory two specimens of *Tribolium castaneum*, the flour beetle were recovered (Table 7.7). As discussed elsewhere, *T. castaneum* is largely associated with flour, and it was the obvious pest to be recovered from the samples. Also, chrysalises of Lepidoptera, possibly *Ephestia* sp. were recovered from the same samples. The lack of pests in the flour could be the result of sieving before storage, and very good storage conditions, or perhaps the grain was processed only shortly before the eruption so that there was little time for any infestation to develop.

Taxon	MN I
Coleoptera	
Tenebrionidae	
Tribolium castaneum (Hbst.)	2
Lepidoptera	
?*Ephestia* sp. chrysalis	few

Table 7-7 Insect fossils from Delta Building

Discussion of the finds

As the archaeoentomological work from other contexts is still in progress, the context for drawing conclusions is the West House. The most useful samples produced a typical storage fauna. Most of the individuals were found infesting the botanical remains, whilst floor samples produced almost no insects. The biogeography of the fauna recovered probably indicates trade links with Egypt and the Near East, and the find of *T. anophthalmus* indicates storage in liquids or as wet pulses. Also, several measures were taken against insect infestation and the use of natural insecticides is likely to have been a common practice. High hygiene standards and a clean environment in general is witnessed by the insect fauna, as no pests of animals or insects related to foul conditions were recovered, outside the latrine.

1. Smothering conditions
by using stone covers and often sealing the mouths of the jars with clay
2. Oils
a. Use of the same pots for storage of crops-liquids
b. Possibly storing in liquid e.g. olive oil (presence of *T. anophtalmus*)
3. Ash (tephra)

Impossible to detect because of the eruption.
4. Minerals
a. Sand (impossible to detect because of the taphonomic problems)
b. Soils (impossible to detect because of the taphonomic problems)
c. Salt -Presence of fish bones in pots
5. Parts of plants
a. *Coriandrum sativum* (seeds)
b. *Echium* sp. (seeds)
c. *Thymelaea* cf. *hirsuta* (leaves)
6. Animal substances
Fish debris (presence of small bones of fish in pithoi)

Table 7-8 : Methods against insect pests practised in the storerooms of the West House, Akrotiri.

Fig. 7. 1. Pithoi set in benches, Beta Building, Akrotiri, Thera.

8. All about silk

Two kinds of silk: true silk and wild silk

Silk thread is obtained by unwinding the cocoons spun by the caterpillars of a number of different species of moth (Lepidoptera). As a fabric, natural silk is the most luxurious material for clothing, desired as much today as in Antiquity. There are different qualities of silk depending on the species of silkworm producing it, as well as on the way the larvae are fed and taken care of. The finest silk is the Chinese one, produced by the Chinese silkworm, the caterpillar of *Bombyx mori* L. From the Byzantine East, *B. mori* was introduced into Europe in the seventh century and since then, its cultivation was spread throughout the world (Scott 1993). However other kinds of silk, inferior to the Chinese, are produced by other lepidopteran species, and their products tend to go under the general name of "wild silk".

The most widely used species of silk moth include *Antherea mylitta* Dr., *A. assama* Westw., *A. pernyi* Guér. & Mén., *A. yamamai* Guér. & Mén., *Philosamia cynthia* Dr., and *P. ricini* Dr. (Zeuner 1963, 484). According to Wardle (1891), there are at least 59 species of silk-producing Lepidoptera (Table 8.1). In the Mediterranean, the wild silk species are *Pachypasa otus*, the largest species of lasiocampid moth in Europe, and *Saturnia pyri,* a saturniid, and the largest moth in Europe (Fig. 8.1).

Today in Assam, wild silkworms are still bred, and the thread from the unravelled cocoons used to produce wild silk garments known as "eri" or "muga" (Gillow & Barnard 1991, 130-1). Although the spread of Chinese silk and industrialization of sericulture has reduced its market, wild silk, "sauvage sois", is still marketed. It is reasonable to assume that in periods when *Bombyx mori* silk was a Chinese monopoly, and an expensive imported luxury, wild silk must have been an alternative material.

Silk's genesis: The Asian mythology

The discovery of silk cocoons signified the beginning of sericulture. A Chinese myth gives the story of the first silk thread: Princess Si Ling-chi, wife of Emperor Huang Ti, of the pre Shang Dynasty, *ca.* mid-third millennium BC, was drinking her tea in the palace garden under a mulberry tree, when a silk worm cocoon fell into her cup. The hot water dissolved the gum of the cocoon and the thread started unwinding. Thereafter a part of the palace garden was dedicated to mulberry trees, and the Empress and her Court ladies gathered the leaves to feed the silkworms. Empress Hsi Ling Shih became the goddess of silk (Barber 1991, 31). The Chinese learnt all the secrets of the native silkmoth, *Bombyx mori*, and started an immensely successful monopoly of silk. Exporting mulberry seeds, for mulberry, *Morus alba,* is what the caterpillar feeds on, and silkworm cocoons was punishable by death, and the secrets of sericulture were well kept for millennia.

In Central Asia, there is a sacred history of the guild of silk weavers. Called "risola", it is more recent than the Chinese myths, and was compiled from legends that originated in oral tradition; it is a mixture of pre-Islamic and Islamic elements. The Bokhara *risola* gives details on silk's genesis:

> "By the order of God, the Archangel Gabriel sent to Job disease, so that his body was covered with worms...but patient Job waited in his place and praised God. There was a command to the Archangel Gabriel to hurl the Prophet Job into the water...and his rotting places became whole. The worms which were in the body of Job crept out into a mulberry tree there and began to eat its leaves, and then knitted themselves up in this covering. This nest where they remained was called, as it is now, pilla, a cocoon."
>
> (Rau 1988, 19).

Fig. 8. 1. The Saturniid moth *Saturnia pyri* and its cocoon (after Zanetti 1980).

The founder of weaving was, according to the *Risola*, Shishi Nabi, "a faceless old man who was born from the seed of Adam, without Eve's help". Another mythological figure associated with the craft was the horned Hazrati Boor, the protector of animal hair weaving. Bibi Rishand is the only female figure in the *Risola* and she is associated with spinning (Rau 1988, 19).

Also in present oral traditions, small innovations and revolutionary ideas concerning the craft are related to the devil. Shaitan is entitled to a part of the earnings of the weaver, the *kha-khash-ast*, but there is no record on how it was paid (*idem*).

Sericulture in the Classical literature

Silkworms and their names

In Greece according to Classical legend, silk cocoon unravelling and weaving was discovered by Pamphile, daughter of Plateus. Silkworms are described in the literature with different names, perhaps partly reflecting the different stages of metamorphosis. It is called σκώληξ (Aristotle, Procopius) or *vermes* (Tertullian, Isidorus) or *vermiculus* (Pliny) (Beavis 1988). Another term used by Aristotle and followed by Clement, Basil and Ambrose is κάμπη, caterpillar, obviously referring to the relevant stage of development of the silkworm (*idem*). The term βομβυλίς (Aristotle) or *bombylis* (Pliny) is used to describe the silk cocoon with the pupa. Similar words, βόμβυξ (Hesychius), and βομβύλιος (Ps. Aristotle) are used for a group of Hymenoptera, while Pliny confuses the Assyrian silkmoth with the Aristotelian Hymenopteran βομβύκιον. However, the word referred to the Hymenoptera comes from the insect's βόμβος, buzzing, while the name referred to the silkmoth has a different origin; perhaps derived from the root "swell", or from a non-Greek dialect. The word βομβύκιον has been associated with the north Syrian commercial city of Bambyke, south of Carchemish (Oppenheim 1967). Perhaps, according to Oppenheim (*idem*), Bambyke was the center for trading the "Assyrian silkworm".

Νεκύδαλος is another name given to the insect by Aristotle, perhaps meaning the adult moth (Forbes 1930, 53). Pliny uses the same term in Latin, *necydallus*, while Clement and Hesychius confuse its meaning. Βόμβυξ or *bombyx*, is the term used more often for silkworms. It is used by Pliny for the Coan silkworm, as well as for the Chinese one by a number of authors: (Hesychius, Clement, Pollux V II.76, Tertullian adv. Marc. I.14, de Pall. III.6, Servius in G. II.121; Isidorus Or. XII.5.8, XIX.22.13, 27.5). It is the standard word in Latin for silkworms: *vermes qui texunt* (V.348.22) or *vermis qui sericum facit*

(IV.602.37). The adult moth is also named by Pliny *papilio* (XI 77-8), while σήρ or *ser* in Latin, meaning Chinese, is used for silk fabrics (Clement: *Julian Ep.* 80), but Pausanias uses it for the silkmoth itself (VI.26.6) (Beavis 1988, 142). In addition, the words used for Chinese silk thread or clothes are derived from the same root σήρ: σηρικά or *serica*.

The Classical sources on wild silk

There is no obvious distinction between wild silk and true silk in the Classical literature. However, knowing that the cultivation of *Bombyx mori* was restricted in China and most probably adjacent areas at least until the 5th century BC, we can safely assume that the early references on silk production from the Mediterranean are records of wild silk cultivation. Both Aristotle and Pliny give us a detailed story of the origin of sericulture in Cos, and a series of Roman and Byzantine writers talk about excellent garments made from natural silk.

Aristotle is the first surviving source to mention wild silk cultivation (HA 452):

> "Ἐκ δέ τινος σκώληκος μεγάλου, ὅς ἔχει οἷον κέρατα καὶ διαφέρει τῶν ἄλλων, γίγνεται πρῶτον μὲν μεταβάλλοντος τοῦ σκώληκος κάμπη, ἔπειτα βομβυλίς, ἐκ δε τούτου νεκύδαλος· ἐν ἓξ δὲ μησὶ μεταβάλλει ταύτας τας μορφὰς πάσας. ἐκ δὲ τούτου τοῦ ζῴου καὶ τὰ βομβύκια ἀναλύουσι τῶν γυναικῶν τινες ἀναπηνιζόμεναι κἄπειτα ὑφαίνουσιν· πρώτη δὲ λέγεται ὑφῆναι ἐν Κῷ Παμφίλη Πλάτεω θυγάτηρ."

> "Out of a certain large larva, which has as it might be horns, and is different from the rest, there arises first of all a caterpillar (produced when the larva metamorphoses) then a *bombylis* (cocoon), then out of this a nekydalos: it goes through all these transformations in six months. Some of the women actually unwind the cocoons from these creatures, by reeling the thread off, and then weave a fabric from it; the first to do this weaving is said to be a woman of Cos named Pamphile, daughter of Plateus."

(trans. Smith & Ross 1910).

From Aristotle's story we cannot identify the species of moth, but it is evident that the species is a native Mediterranean one, and not the Chinese *Bombyx mori*.

Pliny gives a similar story of wild silk cultivation:

> *"Quartum inter haec genus est bombycum, in Assyria proveniens, maius quam supra dicta. nidos luto fingunt salis specie, adplicatos lapidi, tanta duritia ut spiculis perforari vix possint. in his ceras largius quam apes faciunt, dein maiorem vermiculum.*
> *Et alia horum origo. ex grandiore vermiculo gemina protendes sui generis cornua primum urica fit, dein quod vocatur bombylis, ex ea necydallus, ex hoc in sex mensibus bombyx. telas araneorum modo texunt ad vestem luxumque feminarum, quae bombycina appelatur. prima eas redordiri rursusque texere invenit in Coo mulier Pamphile, Plateae filia, non fraudanda gloria excogitatae rationis ut denudet feminas vestis.*
> *Bombycas et in Coo insula nasci tradunt, cupressi, terebinthi, fraxini, quercus florem imbribus decussum terrae halitu animante. fieri autem primo papiliones parvos nudosque, mox frigorum impatienta villis inhorrescere et adversus hiemem tunicas sibi instaurare densas, pedum asperitate radentes foliorum lanuginem, in vellera hanc ab iis cogi subigique unguium carminatione, mox trahi in tramas, tenuari ceu pectine, postea adprehensam corpori involvi nido volubili. tum ab homine tolli fictilibusque in vasis tepore et furfurum esca nutriri, atque ita subnasci sui generis plumas, quibus vestitos ad alia pensa dimitti. quae vero carpta sint lanicia umore lentescere, mox in fila tenuari iunceo fuso. nec puduit has vestes usurpare etiam viros levitatem propter aestivam: in tantum a lorica gerenda discessere mores ut oneri sit etiam vestis. Assyria tamen bombyce adhuc feminis cedimus."*

(HN XI XXV-XVII 75-77)

> "Among these is the fourth genus, the silkmoth (*bombyx*) which occurs in Assyria; it is larger than the kinds (of wasps) mentioned above. Silkmoths make their nests of mud like a sort of salt; they are attached to a stone and are

so hard that can scarcely be pierced with javelins. In these nests they make combs on a larger scale than bees do, and then produce a bigger grub (*vermiculum*). These creatures are also produced in another way. A specially large grub changes into a caterpillar with two projecting horns of a peculiar kind and then into what is called a cocoon (*bombylis*) and this turns into a chrysalis and this in six months into a silk moth. They weave webs like spiders, producing a luxurious material for women's dresses called silk. The process of unravelling these and weaving a thread again was first invented in Cos by a woman named Pamphile, daughter of Plateas, who has the undeniable distinction of having devised a plan to reduce women's clothing to nakedness. Silk moths are also reported to be born in the island of Cos, where vapour out of the ground creates life in the blossom of the cypress, terebinth, ash and oak that has been stripped off by rain. First however, it is said, small butterflies are produced that are bare of down, and then as they cannot endure the cold they grow shaggy tufts of hair and equip themselves with thick jackets against winter, scraping together the down of leaves with the roughness of their feet; this is compressed by them into fleeces and worked over by carding with their claws, and then drawn out into woof-threads, and thinned out as if with a comb, and afterwards taken hold of and wrapped round their body in a coiled nest. Then (they say) they are taken away by man, put in earthenware vessels and reared with warmth and a diet of bran, and so a peculiar kind of feathers sprout out, clad with which they are sent out to other tasks; but tufts of wool plucked off are softened with moisture and then thinned out into threads with a rush spindle. Nor have even men been ashamed to make use of these dresses, because of their lightness in summer: so far have our habits departed from wearing a leather cuirass that even a robe is considered a burden! All the same we so far leave the Assyrian silk-moth to women."

(trans. by Rackham 1940).

Based on Pliny's and Aristotle's story, Forbes (1930) identifies the wild silk worm as *Saturnia pyri* or *Pachypasa otus*. The reasons that made Cos the early centre of sericulture in the Aegean, as well as the actual period when production of silk started there, are problematic, and unfortunately the Classical literature and ethnographic sources provide no other clues.

Amorgos, an island of the Cyclades next to Santorini and near Cos could have also been a centre for silk textiles in Antiquity. Amorgian chitons were very famous during the Classical period. Aristophanes in his comedy Lysistrata talks about the luxurious transparent garments of Amorgos (149-152):

"εἰ γὰρ καθοίμεθ' ἔνδον
ἐντετριμμέναι
κἀν τοῖς χιτωνίοισι τοῖς ἀμοργίνοις
γυμναὶ παρίοιμεν, δέλτα
παρατετιλμέναι,
στύοιντο δ' ἄνδρες κἀπιθυμοῖεν
πλεκοῦν,..."

"For if we (women) will but sit at home,
powdered and trimmed, clad in our
daintiest lawn
employing all our charms, and our arts
to win men's love..."

(trans. by Rogers 1924).

and also (43-45):

"αἵ καθήμεθ' ἐξηνθισμέναι,
κροκωτά φοροῦσαι καί
κεκαλλωπισμέναι
καὶ Κιμβερίκ' ὀρθοστάδια καί
περιβαρίδας,.."

"we who sit trimmed
and bedizened in our saffron silks,
our cambric robes, and little finical
shoes"

(trans. by Rogers (1924)).

When one of Lysistrata's friends tries to escape, she uses as an excuse (735-736):

" τάλαινα ἐγώ, τάλαινα τῆς
ἀμοργίδος,
ἥν ἄλοπον οἶκοι καταλέλοιφ'."

"O goodness gracious! O that lovely
flax I left at home unhackled."

(trans. by Rogers 1924).

And Lysistrata answers (737-738):

"αὔτη 'τέρα.

ἐπί τὴν ἄμοργιν τὴν ἄλοπον ἐξέρχεται."

"Here's another. She's stealing off to hackle flax forsooth."

(trans. by Rogers 1924).

The translator refers to *Amorgis* as flax while others (e.g. Forbes 1956; Beavis 1988), suggest that it should be translated as silk cocoons, silk.

Plato (*Epistles* XIII, 363A) writes also about Amorgian chitons:

"Καὶ ταῖς Κέβητος θυγατράσι χιτώνια τρία επτάπηχη, μὴ τῶν πολυτελῶν τῶν Αμοργινῶν, ἀλλά τῶν Σικελικῶν τῶν λινῶν."

"and to the daughters of Cebes three tunics of seven cubits, not made of the costly Amorgos stuff but of Sicilian linen."

(trans. by Bury 1929).

In Suda it is explained that

Σ1625.
"Ἀμοργινόν· ὅμοιον βύσσω καὶ πολυτελὲς...
.....ἔστι δὲ ἡ ἀμοργίς ὅμοιον ἀλεπίστω λίνω. περιλεπίζουσι δὲ αὐτῷ καὶ ἐργάζονται. ἔστι δὲ σφόδρα λεπτό ὑπὸ τὴν βύσσον ἢ τὴν κάρπασον."

"Amorginon is like linen and very dear...Amorgis is like unhackled linen, they strip it and work it. It is much finer than cotton (byssos) or linen (karpasos)."

Richter (1929) believed that the flowing garments on fifth century statues are "Amorgian chitons", the best quality clothes during that era, honoured offerings to the goddess Artemis. Forbes (1956) believed that cocoons of wild silk may even have been imported from the East, where the "Assyrian silk moth " mentioned by Pliny is native. However, it can be argued that wild silk production utilised a Mediterranean moth species, and could have been practised in the Aegean long before the introduction of Chinese silk.

Herodotus talks about silk clothes, naming them Median (I, 135 and II, 84):

"καὶ γὰρ δὴ τὴν Μηδικὴν ἐσθῆτα νομίσαντες τῆς ἑωυτῶν εἶναι καλλίω φορέουσι, καὶ ἐς τοὺς πολέμους τοὺς Αἰγυπτίους θώρακας."

"They wear the Median dress, deeming it more beautiful than their own, and the Egyptian cuirass in war."

(trans. by Godley 1926),

and

"..ἐξαίρετα διδόσθαι ἐσθῆτα τέ Μηδικὴν ἐτέος ἑκάστου."

"...should receive for themselves specially a yearly gift of Median raiment."

(trans. by Godley 1926).

Xerxes also, according to Herodotus (VII, 116), declared the Acanthians friends and gave them a present of Median dress.

Xenophon talks about the Median dress (*Cyropaedia* VII i 40):

"στολήν τε γοῦν εἵλετο τήν Μηδικήν αὐτός τε φορεῖν καὶ τούς κοινῶνας ταύτην ἔπεισεν ἐνδύεσθαι."

At any rate, he chose to wear the Median dress himself and persuaded his associates also to adopt it.

(trans. by Miller 1914).

Most probably the "Median dress" was made of wild silk from India, or even silk produced by *B. mori* imported to India through the silk route.

Tertullian, a Christian writer, describes Alexander's fascination for Median dress (*De Pall.* 4):

"Vicerat Medicam gentem, et victus est Medica veste. Triumphalem cataphracten amolitus in captiva sarabara decessit; pectus squamarum signculis disculptum textu pellucido tegendo nudavit, ahhelum adhunc ab opere belli, et ut mollius, ventilare serico extinxit. Non erat satis animi tumens Macedo, ni illum etiam vestis inflatior delectasset."

"He had conquered the Median race, and was conquered by Median garb. Doffing the triumphal mail, he degraded himself into the captive trousers! The

breast dissculptured with scaly bosses,
by covering it with a transparent texture
he bared; panting still after the work of
war, and (as it were) softening, he
extinguished it with the ventilating silk!"

((trans.) Tertullian 1895).

Procopius later on testifies that the Median dress was silk (*History of the wars* I, 20. 9-12):

"αὕτη δέ ἐστιν ἡ μέταξα ἐξ ἧς εἰώθασι τὴν ἐσθῆτα ἐργάζεσθαι ἥν πάλαι μὲν Ἕλληνες Μηδικὴν ἐκάλουν, τανῦν δὲ σηρικὴν ὀνομάζουσιν."

"This is the silk of which they are accustomed to make the garments which of old the Greeks called Medic but which at the present time they name "seric"."

(trans. by Havry 1905).

The Periplus of the Erythraean Sea, a compilation of an Anonymous author, *ca.* 100 AD, is an excellent source of information on trade in the Eastern Roman Empire at the time. Apparently northwest (port of *Barbaricon* 39:13.11, port of *Barygaza* 49:16.30), southwest and southern India (*Nelkynda* 56: 18.24), and also northeast India (*Ganges* 64: 21.13-14) were exporting silk cloth, yarn and silk floss (Casson 1989). The word translated as silk floss is the Greek word used for wool. Wild (1984) believes that it could also mean wild Indian silk. It is possible, however, that the silk came from China, and India was the intermediate post to Europe.

Horace, born around 65 BC and a master of the lyric form, describes a seductive outfit, a dress that does not "hinder you from clear view." (*Satires* I ii 101-103):

"Cois tibi paene videre est ut nudam, ne crure malo, ne sit pede turpi; metiri possis oculo latus."

"In her Coan silk you may see her, almost as if naked, so that she may not have a poor leg, an unsightly foot; you may measure her whole form with your eye."

(trans. by Fairclough 1966).

Interweaving with gold threads was something common in Coan garments, as witnessed by Tibullus (II iii 53-4), also a poet, who died in 19 BC.

"illa gerat vestes tenues, quas femina Coa texuit, auratas disposuitque vias,"

"Let her wear the gossamer robe which some woman of Cos has woven and laid it out in golden tracks."

(trans. by Postgate 1988).

Clothes created in Cos were so celebrated that apparently they were the equivalent of a fashion label of today. Ovid writes in the *Ars Amatoria* (II 297-298), written in the late first century BC:

"Sive erit in Tyriis, Tyrios laudabis amictus: Sive erit in Cois, Coa decere puta."

"If she be in Tyrian attire, then praise her Tyrian gown; or in Coan, then find the Coan style becoming."

(trans. by Mozley 1939).

The Byzantine writer, Ioannes Lydus explains further why Coan clothes were so unique:

"μανδύην μὲν γὰρ ὁ ἔπαρχος περιεβάλλετο Κῷον (ἐπ' ἐκείνης γὰρ τῆς νήσου καὶ μόνης ἡ βαθυτέρα βαφὴ τοῦ φοινικοῦ χρώματος τὸ πρὶν ἐπηνεῖτο κατασκευαζομένη· (ἡ) γὰρ ἠρέμα πως ἐπὶ τὸ φλόγινον καὶ οὐ σφόδρα βαθὺ αναπτομένη πρὸς Παρθυαίων εξηύρηται· ὅθεν καὶ Παρθικὰ τὰ φλογοβαφῆ δέρματα συμβαίνει καλεῖσθαι)·"

"For the prefect to wear a Coan cloak, for the rather deep dye of crimson hue which was prepared on that island and that alone was extolled in former times; for the dye which is tinctured somewhat gently towards the colour of flame and is not exceedingly deep has been discovered by the Parthians; for this reason actually flame coloured hides happen to be called "Parthic"."

(trans. by Bandy 1983).

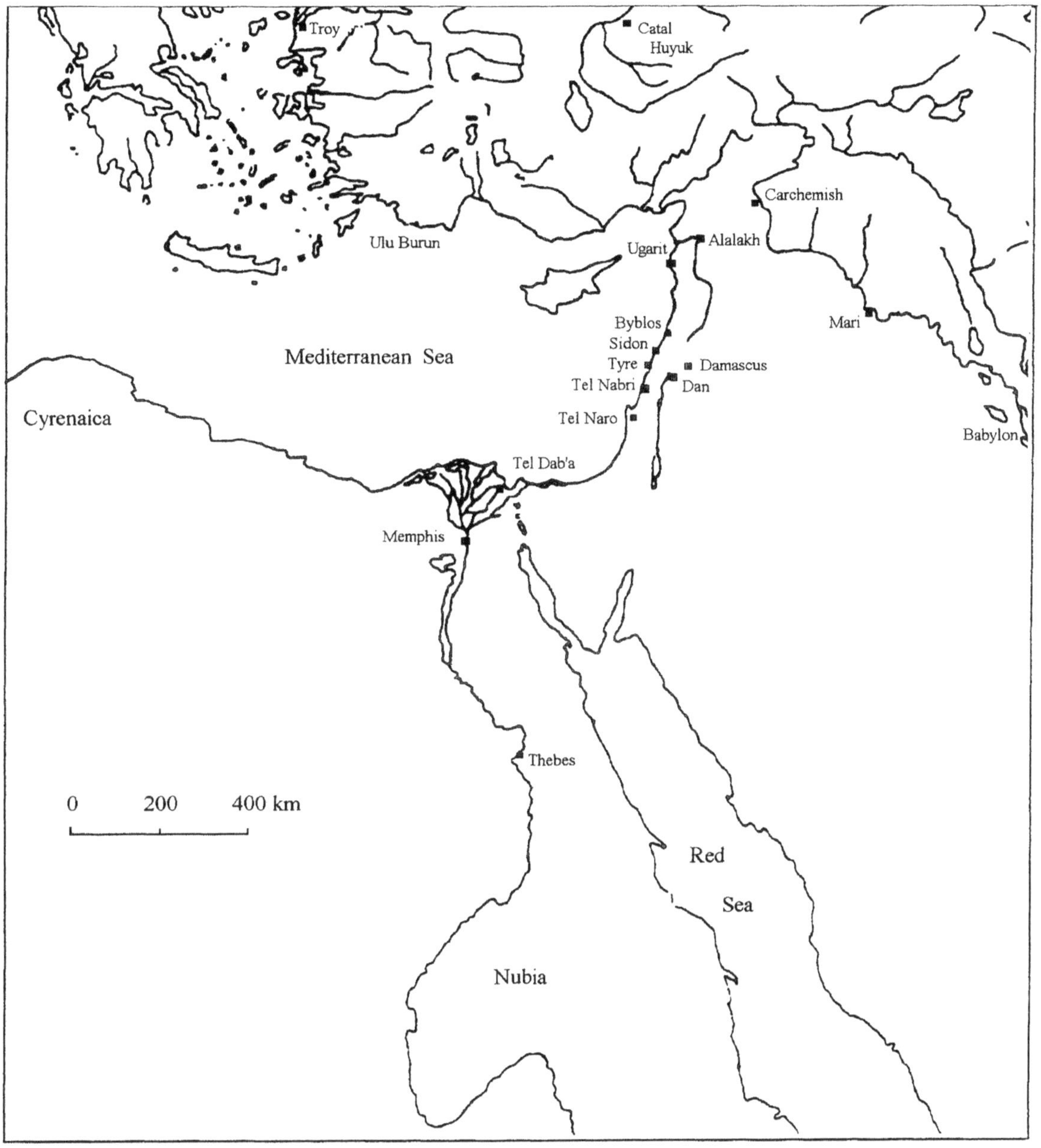

Fig. 8. 2. Sites in the Eastern Mediterranean region mentioned in the text.

The texture and the deep purple colour of Coan garments may have created a whole fashion, just as the Amorgan chitons during the third century BC in Athens were unique and desirable luxury items, because of their deep red colour and fineness. As already noted, their names were more a guarantee of quality material than a reference to the place they were made. It is apparent that wild silk workshops existed in the Aegean during Classical times. Most probably the islands of Cos and Amorgos controlled wild silk production and trade, and set the standards for the market in wild silk.

Procopius in his biography of Justinian refers to (*Anec.* 25.14-15) manufacture of silk from ancient times at Berytus and Sidon (Fig. 8.2). Propertius also refers to *Arabius bombyx* (II 3 15) and Lucan to *Sidonium filum* (X.141). It is probable that other species and not *Bombyx mori* are recorded in these passages (Beavis 1988), since it would be naive to assume that sericulture and the true silk textile industry was born overnight. Probably it came to take over a market that could appreciate high quality garments, and knew already all about silk in the form of rougher, wild silk. So, it comes as no surprise that, during the reign of Justinian, Chinese silk started to dominate the market.

Chinese silk and its introduction to Europe in the Classical record

Pliny describes another material, wrongly confused with silk (Forbes 1956) that came from *Seres*:

> *"Seres, lanicio silvarum nobiles, perfusam aqua depectentes frondium canitiem, unde geminus feminis nostris labos redordiendi fila rursuque texendi."*
>
> (HN VI LIV)

> "The first human occupants are the people called the Chinese, who are famous for the woollen substance obtained from their forests; after a soaking in water they comb off the white down of the leaves, and so supply our women with double task of unravelling the threads and weaving them together again."
>
> (trans. by Rackham 1942).

Most probably what he describes in the above passage is bast fibre and not silk. Confusion is also evident in the "Geography" of Strabo (15 1 20). Nearchus, after talking about cotton, gives information on "shrika", silk clothes:

> "τοιαῦτα δέ καὶ τὰ Σηρικά, ἐκ τινῶν φλοιῶν ξαινομένης βύσσου."
>
> "The serica also are of this kind, byssus being dried out of certain barks."
>
> (trans. by Jones 1930).

In the case of Nearchus, the wrong concept of silk could be attributed to either lack of information or misinformation by the Chinese. Virgil (Georg. II.121) characterises the *Seres*, Chinese, as those who "*vellera...foliis depectant tenuia*". Seneca, (Hipp. 389, Herc. Oet. 666-7), Solinus (50.2), Ammianus Marcellinus (XXIII.6.67), Dionysius Periegetes (Orb. Descr. 752 ff.), Eustathius (Comm), Claudian (I.179-80), Ausonius (Tech X.24), Martianus (Capella VI.693) and Isidorus (Or. XIV.3.29) all believed in a plant origin for silk (Beavis 1988).

Pausanias (VI 26.6-9) gives the first detailed description of the Chinese silkworm. However, his information is not correct, most probably because he obtained it from traders or an embassy of Marcus Aurelius to China in 166 AD recorded in Chinese sources (Frazer 1913); the Chinese may have been deliberately misinforming their guests. He believed that silkworms were fed on millet and fresh reed for the last year, and that their breeding period was five years. Afterwards, he writes, the silkworms burst open and died and most of the thread was found inside them. Pausanias' informants may have been misinformed, but a useful product can be made from the glands of dead silkworms. For this product, the life cycle of the silkworm has to be interrupted at the end of the larval stage. The silk glands are extracted and stretched as long as eighteen inches and as thick as heavy thread. The pieces of gut can be used for fishing as attachments at the end of a fishing line or in surgery. Murcia in Spain was said to specialise in fine silkworm gut (Cooper 1962, 79-84).

Procopius of Caesarea (VIII.17) gives the story of the monks who offered to bring silkmoths and breed them for the Byzantine emperor.

> "About the same time certain monks arrived from the (country of the) Indians, and learning that the Emperor Justinian had it much at heart that the Romans should no longer buy silk from the Persians, they came to the king and promised that the Romans should no longer buy silk from the Persians or from any other nation; for they had lived, they said, a long time in the country where there were many nations of the Indians, and which goes by the name Serinda. When there they had made themselves thoroughly acquainted with the way in which silk might be produced in the Roman territory. When the Emperor questioned them very closely and asked how they could guarantee success in the business, the monks told him that the agents in the production of silk were certain caterpillars, working under nature's teaching, which continually urged them to their task. To bring live caterpillars from that country would be impracticable indeed, but arrangements might be made for hatching them easily and expeditiously. For the eggs produced at birth by one of those worms were innumerable and it was possible to hatch these eggs long after they had been laid by covering them with dung which produced sufficient heat for the purpose. When they had given these

the Persians, they came to the king and promised that the Romans should no longer buy silk from the Persians or from any other nation; for they had lived, they said, a long time in the country where there were many nations of the Indians, and which goes by the name Serinda. When there they had made themselves thoroughly acquainted with the way in which silk might be produced in the Roman territory. When the Emperor questioned them very closely and asked how they could guarantee success in the business, the monks told him that the agents in the production of silk were certain caterpillars, working under nature's teaching, which continually urged them to their task. To bring live caterpillars from that country would be impracticable indeed, but arrangements might be made for hatching them easily and expeditiously. For the eggs produced at birth by one of those worms were innumerable and it was possible to hatch these eggs long after they had been laid by covering them with dung which produced sufficient heat for the purpose. When they had given these explanations, the emperor made them large promises of reward if they could verify their assertions by carrying the thing into execution. So they went back again to India and brought a supply of the eggs to Byzantium. Having treated them just as they had said, they succeeded in developing the caterpillars, which they fed upon the mulberry leaves. From this beginning originated the establishment of silk-culture in the Roman territory."

(trans. by Havry 1905).

Theophanes also (*Photius Bibl.*64) describes the transformation of the larvae to moths.

"Now in the reign of Justinian, a certain Persian exhibited in Byzantium the mode in which (silk) worms are hatched, a thing which the Romans had never known before. The Persian on coming away from the country of the Seres had taken with him the eggs of these worms (concealed) in a walking stick, and succeeded in bringing them safely to Byzantium. In the beginning of spring he put out the eggs upon the mulberry leaves which form their food and the worms feeding upon those leaves developed into winged insects and performed their other operations. Afterwards when the Emperor Justinian showed the Turks the manner in which the worms were hatched, and the silk which they produced, he astonished them greatly. For at that time the Turks were in possession of the marts and ports frequented by the Seres, which had been formerly in the possession of the Persians. For when Ephthalanus, King of the Ephthalites (from which indeed the race derived that name), conquered Perozes and the Persians, these latter were deprived of those places, and the Ephthalites became possessed of them. But somewhat later the Turks again conquered the Ephthalites and took the places from them in turn."

However, a Chinese document, *San-kuo-chih*, refers to Ta-tsin (later Fulin identified as Syria), and gives a list of its domestic animals: "the donkey, the camel and the mulberry silkworm" (Hirth 1966). Knowing that San-kuo-chih was compiled before 429 AD, Muthesius (1993, 22-23) assumes that sericulture was introduced earlier than the sixth century AD in Byzantium, from Syria and not China. Also, the reference to mulberry silkworm can only mean that other kinds of silkworms - wild ones, non-mulberry feeders - must have been known and exploited outside China at that period.

Muthesius (1993, 22) further suggests therefore that cultivation of *B. mori* might have already started during the fifth century, and the stories of the historians could be a tribute to Justinian for establishing the true silk industry in Constantinople itself.

Silk in the Bronze Age Mediterranean

As no standard name exists in the literary record for true or wild silk until the widespread use of the term *sericon*, it is apparent that silk was described in other ways, related to its quality, value, area of production or manufacture. There are a number of Linear B tablets (Fig. 8.3) that refer to textiles (Ventris & Chadwick (1973; Killen 1966). These are characterized by Evans' 'banner' ideogram, which probably represents an upright loom with loom-weights on the warp. The most characteristic term is *pa-we-a*, meaning *pharwea*, a piece of cloth (*cf.* Homeric *faros* = a large cloth for a sail, a large cloak, a funerary shroud).

The ideogram translated as wool is derived from Linear A and the weight of wool is counted in *M* units. Bennett (1950) believed that M3 was the normal wool unit of weight, equivalent to about 3 kg, whilst Sundwall (1932) suggested that the Wool ideogram is exclusively a unit of value but, in -D, a special Cretan group of tablets referring mainly to textiles, there are several problematic points. For example in the Lc525 tablet the following is written:

se-to-ja: wa-na-ka-te-ra cloth+TE 40 wool 200 [+ *tu-na-no* CLOTH 3 WOOL [nn]

From Se-to-ja : Forty edged cloths of royal type, 200+ measures of wool; Three cloths of tu-na-no type, several hundred measures of wool.

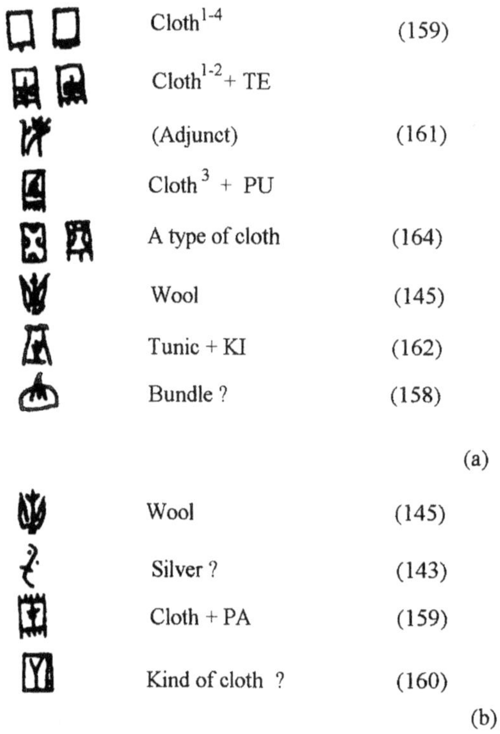

Fig. 8. 3 : Linear B ideograms from (a) Knossos and (b) Pylos (redrawn from Ventris & Chadwick 1973).

In tablet Lc532 the tu-na-no entries give 3 measures of wool per cloth and in Lc530 4 measures per cloth, while in Lc525 it is about 100 measures. The amounts of wool per Cloth+te give 6.5 measures in Lc532, 7 measures in Le553, 6 in L520 and 5 in Lc525. These discrepancies have led to questioning of the use of the Wool ideogram as a unit for value; indeed Killen (1966) believes it to be a measure of wool given out to groups of women to make a certain type of cloth. In this way, the variable amount would reflect textiles of different types which require varying amounts of wool for their completion. If it is to be believed, however, that in some instances the wool is used as an equivalent of value, then the *tu-na-no* type cloths in Lc525 could be explained as being of great value; one possible candidate is silk.

Turning to another inscribed tablet, J693:

ri-no/re-po-to 'qe-te-o' ki-to BRONZE I [*sa-pa* 2 I *e-pi-ki-to-ni-ja* BRONZE I

Fine linen, of the tribute: a tunic = 1kg. of bronze...a sa-pa = 45g. (of bronze). over-shirt(s)=1 kg. of bronze...

In this tablet the fine linen is related to bronze units instead of wool units. Here there is no possibility of an explanation of bronze being material provided centrally to produce the whole craft item, nor does it seem to be an equivalent weight, as it seems improbable that a fine linen tunic would weigh 1kg. The bronze may be used here as a measure of value. Barber (1991, 313) argues that the bronze was to be used in the decoration of cloth by attachments, whereas Ventris and Chadwick (1973) suggest the use of metal and textile to make armour. This still leaves the term "fine linen" to be explained. If the value of the tunic, overshirts and *'sa-pa'* is so high, it is justified to assume that they have been made of a very valuable fabric. Therefore it is suggested that the term *ri-no/re-po-to* in Linear B (fine linen) might represent wild silk (as already mentioned, Suda describing "'Αμοργινόν" or wild silk, according to Forbes (1956), and Richter (1929), suggests that it "is like linen and very expensive", and concerning ἀμοργίς, "it is much finer than cotton or linen".

Such a connection may be strengthened by comparative evidence. The measurement of the value of fine linen might not be fortuitous, linked as it is to bronze and tablets which list armour. Silk tunics form a natural part of body armour in combination with metal, as they minimise the impact of arrows etc., allowing their removal, as the point of the projectile does not pierce the silk cloth. It is for this reason that the Mongols used silk undergarments with body armour (Marshall 1992, 39-40) The term *ki-to* meaning khiton, tunic, or possibly fine linen, in Akkadian *kittintu*, Ugaritic *ktn*, Hebrew *k[e]tonet*, may derive from Mesopotamia (Ventris & Chadwick 1973, 554; Duhoux 1988, 79). *Kito* is associated with the armour and the corslet and, according to Oppenheim (1967), it could imply garments made of wild silk. If the above is a reference to a costume made of silk to be worn together with the armour, it can be valued in the same way as the armour itself, in bronze units.

In L 587+589+596 (G) we read the following:

> po-ki-ro-nu-ka CLOTH 24 re-u-ko-nu-ka CLOTH 372
> ko-ro-ta CLOTH 14 pa-ra-ku-ja CLOTH 42 po-ri-wa CLOTH I
> edge: to-sa CLOTH 149
>
> Twenty four cloths with coloured o-nu-ke, 372 with white o-nuke, fourteen dyed cloths, forty two of the colour of pa-ra-ku, one grey one.

Ko-ro-ta could be *krokia*, dyed with crocus, yellow (Fig. 8.4). Pa-ra-ku-ja was either dyed or decorated with *paraku*. *Pa-ra-ku* is probably a word referring to precious material, but it is not certain to what exactly. It is identified as silver (*+falargys*, *parargys*: next to silver) or as emerald ((from the Akkadian word *barakdu*). Maybe here *pa-ra-ku-ja* are not the cloths coloured with *paraku*, but the clothes woven with silver threads, or decorated with silver beads. Decoration with gold and silver beads was often related with silk; the precious metals were used to cover the fine, but extremely strong and durable precious silk threads.

Also *o-nu-ke*, nails, re-u-ko-nu-ka for clothes decorated with white "nails" and po-ki-lo-nu-ka for decoration with "nails" of varied colours are terms used in Linear B in relation to textiles. In Soufli, Thrace, Greece, cloths are still named by the patterns woven on them, and this may have been the case with *reukonuka* and *pokironuka* garments (Oikonomou 1992, 98).

Also problematic is the use of the word *arioa* in Ld571 tablet:

> pe-ne-we-ta a-ro-a BUNDLE? I
> pa-we-a/e-qe-si-ja re-u-ko-nu-ka
> CLOTH 25
>
> Twenty five cloaks with white o-nu-ke suitable for followers, provided with pe-ne-, of better quality; and one bundle.

Killen notes that the term *arioa* was never used for the *xenwia* (for gifts, or for export) fabric (Killen 1985, 263, 292 n.66). We have already found quotations referring to wa-na-ka-te-ro (for the king) and e-qe-si-ja (for the followers). Whether these terms imply better quality wool, elaborate weaving, or the use of materials such as silk or cotton cannot be proved.

The monopoly of silk

> "Cleopatra's white breasts resplendent through the Sidonian fabric, which wrought in close texture by the skill of the Seres, the needle of the workmen of the Nile has separated, and has loosened the warp by stretching out the web."
>
> Lucan, *Pharsalia* 141.

The Chinese guarded their monopoly of true silk until the reign of Justinian. Silk garments were sent overland, from Turkestan through Persia, India, and the desert, to Syria and the Mediterranean, and by sea from India and Ceylon to Alexandria, or from Central Asia to the Caspian and overland to the Black Sea (Wild 1984). Silk was very expensive and restricted to the wealthiest nobles only. During Tiberius' reign there was a law discouraging men from wearing, "the effeminate delicacy of silk apparel"

(Tacitus, *Annals* II, 33.1).

Aurelian (AD 270-275) refused to wear it (*Scr. Hist. Aug.* Severus Alexander 40), and did not allow his wife to buy clothes that cost their weight in gold. Diocletian (284-305) in his Edict (Frank 1940, 373-378) notes only small quantities of silk or silk interwoven with other threads. Theodosius in the *Codex* claims it a restricted luxury, especially when interwoven with gold. The Theodosian Code, the Justinian Code and, a few centuries later, the Basilics give evidence of silk legislation over a period of six centuries. The main issues were:

- maintenance of an Imperial silk weaving and purple dyeing force
- prevention of fraud within the imperial workshops,
- protection of the Imperial monopoly over murex purple and certain gold borders.

(Muthesius 1993, 57).

When finally the breeding of *Bombyx mori* was established, a number of stricter laws ensured a Byzantine monopoly on silk trade. Leo VI in his *Book of the Prefect* forbade *serikarioi* (clothiers and dyers of silk) to dye silk purple, and make the "forbidden clothes" (Muthesius 1993, 47). The Byzantine monopoly was well kept till the eleventh century AD, when the emperors started giving away to the barbarians rights on the silk trade in an attempt to ensure the safety of the Empire.

Wild silk and Byzantium

There is no mention of wild silk in the Byzantine records. The Theodosian code and the Justinian code issued regulations about silk principally in Constantinople, and this may imply that only a very limited amount of true silk was imported and woven in Imperial workshops. However, there had been a long tradition of luxurious garments from early Antiquity, and it is reasonable to assume that the wild silk workshops were still supplying the Empire with fine textiles. It is probable that long history and prestige allowed them to keep their independence. As Jones (1986), notes the supply of silk clothing must have been heavily dependent upon the private sector. It is worth mentioning a passage from the price edict of Diocletian (Frank 1940), where the relative costs of treating the silk yarn are documented: for spinning first quality purple silk, workers earned up to 116 denarii an ounce and purple silk of second quality was spun for 60 denarii an ounce. We need to clarify that silk was dyed with the most costly kind of purple, so the quality distinction refers to the silk itself and not to the dye. It would be unwise to think that "second quality" silk means silk from the outer part of the cocoon or from hatched cocoons or even from inferior cultures of *B. mori*, because it would not have been dyed with the most expensive colour. Most probably it refers to wild silk as opposed to the "first quality", imported and far more exclusive true silk.

Another interesting point is the price of silk during Justinian's reign. It was the period that κομμερσάριοι, merchants with Syria, mainly dealing with silk, appear for the first time. A law issued by Justinian set a ceiling price of 8 nomismata for 1 libre weight of raw silk. However another law, an undated Novella, set the ceiling price of raw silk bought by the κομμερσάριοι, at 15 nomismata a libre. Lopez (1945) was convinced that the Novella was of a post-Justinian date. Oikonomides (1986) argued for an 8 nomismata ceiling price for light silk garments, as opposed to a whole pound of silk. Muthesius (1993, 24) thought that Justinian, after setting the law for 8 nomismata per pound, changed his mind because the low price brought a drastic shortage of raw material. But it could also be that the first price was set for local silk, wild silk produced in the Mediterranean textile centres, and the second for Syrian true silk.

Sericulture

Breeding silkworms

Sericulture is an activity still practised in some places in the traditional way in Middle Eastern countries, Central Asia and the Mediterranean. Different species of silk worms have different eating habits, and give different qualities and colours of silk (e.g. white, yellow, brown, silver grey). *Bombyx* species feed specifically on mulberry leaves. In particular, *Morus alba* is considered the appropriate feed because silkworms which breed on it give the best quality silk. The other species of mulberry are less good for sericulture as the silk produced is inferior (Wardle 1904, 32; Sarpaki, pers. comm.).

The process of feeding the silkworms is very time consuming and is mainly a task undertaken by women. In the Greek village of Soufli, Thrace (Oikonomou 1992, 72-99), the young silkworms are put on mulberry branches placed on a special type of bed (made of wood tiles and covered with paper), while Thai women prefer to feed them in shallow trays covered with paper. The paper is changed daily in order to avoid fungal infections of the silkworms. Thai women also place the trays over pots with water, and cover them with cotton or old silk clothes so as to avoid infestation by insect pests, flies and wasps.

The breeding cycle lasts around a month. The silkworm, during this time, passes through five different "instars" and four stages of "rest". The first instar lasts for four and a half days, and the silkworms have to be fed every two hours, according to the ethnographic evidence from Soufli (Oikonomou 1992, 72-99). The species used in Thailand is fed three or four times a day. The mulberry leaves have to be clean, fresh, and finely chopped. Mulberry leaves used for feeding on Indian silkfarms, are thoroughly cleaned and dried from excess water (S. Anderson, pers. comm.). In Thailand, they are wrapped in banana leaves so as to avoid drying out. As mentioned previously, during all stages, it is essential for the surroundings of the silkworms to be kept clean, as they are very susceptible to fungal attack, infestation and disease. The first rest stage lasts one day: they stay immobile and "change skin", as the people of the village say. Afterwards they are taken to new beds and they are fed every two hours exactly as during the first instar. The second instar lasts four days, the third instar's duration is four to five days, and then follows a fourth and a fifth instar of approximately similar duration. During the 48 hours after the fifth instar, they search for a place to make the cocoon. They climb on a branch and get attached to it . It takes about 50 hours for the larva to be closed entirely inside the silk cocoon. The women of Soufli in Greece take the cocoons off the branches during the next three days and keep them in a cool and clean place. The next step is to kill the moth before it emerges and breaks the thread. This happens after fifteen days, according to the Greek ethnographic record (Oikonomou *idem*), or ten days in Thai sericulture (Conway 1992, 66).

Killing the "butterflies"

There are several ways used traditionally for killing the moths before they hatch from the cocoons. In Soufli, they put the cocoons in a pot that is placed over boiling water and steamed for ten minutes. Then they open one to test if the moth is dead. After ensuring that the moths are killed, the cocoons are dried and put in a clean place. Another method that used to be practised in the village was baking the cocoons. In Thai villages, the women put the cocoons straight into boiling water and leave them for some minutes. In Afghanistan, they just expose them under the hot sun. Apparently the last method is the most risky one, as sometimes moths do not get killed and they destroy the cocoons as they emerge. In medieval China, in order to delay the emergence of the moths, they used to store the cocoons in salted jars. For some reason this treatment was said to make silk brittle (Muthesius 1993, 33).

Reeling the silk thread

The thread produced from the outside of the cocoon is inferior to the silk thread produced from the filament of the cocoon. The process of reeling the silk thread is very laborious. The necessary items for this process are: a boiler (a metal cauldron or a clay pot), a wooden wheel moved by hand, around which the silk thread is wound, and other implements that vary in different cultures: a small broom with which the cocoons were caught and a pan with holes made using nails that served to take out the cocoons from water in Soufli, or a forked bamboo batten in Thailand.

In Soufli they also put opposite the wall two parallel pieces of wood. In one of these there is a hole through which the thread passes. Two bobbins are attached on the pieces of wood and this way the thread is reeled more easily and does not get entangled. The silk thread has to be of a certain thickness and some times twenty cocoons are used to make one single thread. In Thai villages during the process of reeling, the women twist the thread with their fingers. They say that this is the best quality thread, rounded evenly and shiny, and it dyes far better than industrial threads (Conway 1992).
As for the silkworms, they are cooked with chili, garlic, salt and shallots, or roasted in banana leaves with herbs and spices and served as a special delicacy in Thailand. They are the favourite snacks of children who stand near the cauldron and wait to choose the biggest ones (Conway 1992, 68). In Afghan villages, they are used as fertilisers for gardens and fields (Reut 1983, 41).

On wild silk moths

The ethnographic information available (Wardle 1891), is about tussur silk, produced by *Antherea mylitta,* the most widely known Indian wild silkmoth, and *A. pernyi,* its Chinese counterpart. The Indian moth is distinguished by four large, transparent wing spots, surrounded by a blue circle, larger in the female than in the male. The Indians believed that they are the chakra or discus of the god Vishnu, or that they were caused by the touch of his figures, and the moth was considered a sacred insect (Wardle 1891, 3-4). The cocoons of the Indian moth are hard, well formed, shell-like and shaped like a pullet's egg, and very compact. They are variable in size (from 1.8-25 mm long and about 1 and 3 mm in diameter). They possess pedicular attachments, and their very hard, cement like cocoons are attached to the branches in a fruit-like manner, during metamorphosis, which can last for up to six months. Because of natural hardness, the cocoon can only be opened with a penknife and, with its strong pedicular attachment, this aids survival from predation in the wild. When the chrysalis is ready to emerge, an acid fluid is released that breaks the cocoon and enables the insect to emerge (*idem*, 6-7).

An interesting account of the Santals, an Indian grouping of seven tribes, who based part of their subsistence on the exploitation of the tussur silkmoth, is given by Wardle (1891, 20-23). When the tussur moths were at their last stage, the Santals picked them from the branches of the trees, mainly *Terminalia tomentosa* and *Shorea robusta*, by cutting the branches with hooked knives and long sticks. In captivity, the female moths were put in a basket so that all the eggs were gathered, but, in the wild, the eggs were left on the leaves (*idem*). In cultivating the tussur moth in a semi-domesticated way, it was fed on gathered leaves. The eggs were taken into the jungle in small bags and attached to the branches and then they were watched for insect predators night and day for a week. *Semecapens anacardium* was used on a stick, or a band made around the stem of the tree with oil of *Semecapens*, so as to prevent ants and other insect pests (*idem*).

The religion of the Santals was related to tussur moth culture. The Santals believed in evil spirits and in a god that punished them. They had clay altars and sacred groves where they went to complain to the gods and ancestors for their misfortunes, and have been seen striking the altars. They had in every village a sacred grove, consisting of about a dozen large *Sal* or *Shorea robusta*, the favorite tree of the silkmoth, and seven stones, some of them marked red. It was believed among the Santals that the grove was the resort of their deified spirits and ancestral ghosts, and there they would perform all their festivities and rituals (Wardle 1891, 20).

Preparing the thread for weaving

Reeling is the process of lightly twisting long threads. Spinning is the process of twisting and drawing out the plant or animal material into a thread. Silk thread is mostly reeled and not spun. In Thai villages it is twisted lightly by using the fingers and, rarely for short fibres (from the outside of the cocoon or from hatched cocoons), it is spun using a spindle. Very fine and transparent woven clothes were either unspun or very lightly spun in Soufli. However, on very special occasions, they spun silk; the thread used as weft was always the better spun.

Before weaving in Greece, they used to measure the thread on a long wooden stick (τυλιγάδι). One measurement was 60-63 threads of 60 to 80 cm each. Firstly the glue had to come out of the threads. This was done by boiling the silk thread. Then the thread was dyed or boiled (in water with cow dung in Soufli) to get its original colour (Oikonomou 1992, 95-96). Also, in India wild silk was softened with *sagi-mutta*, an impure carbonate of soda to which cow dung had been added. The thread was fixed in an astringent soapy decoction of the nut of *Terminalia chebula* or *T. myrobalana* (Wardle 1891, 25). In Thailand, they prepare a solution of banana tree trunk ash or spinach leaf ash with water, and, after boiling, straining and cooling it, they boiled the silk skeins in it so as to de-gum them.

Afterwards, they prepare the skeins for dyeing in liquid made from thorny vine root and water. During all these processes, it was easier for the workers if the silk was in skeins. In Thailand, women used to wind the thread around small bamboo skeins so that the hank of thread could fit afterwards into four pint pots used for dying. Afterwards, they would wet the skeins of yarn, hang them on sticks, and then put them in the plant dye.

Early in the present millennium (Peroy 1963, 84-90 in Muthesius 1991) in France, among other people occupied with sericulture and silk weaving, there was a special group of women (devideresses) who used to rewind the already spun yarn into the desired skeins or bobbins. In Byzantine times, καταρτάριοι were a separate guild probably specialising in unravelling and reeling the silk as well as degumming it (Muthesius 1993, 32).

Dyeing

Dyeing silk yarn and cloth (Fig. 8.5) is a technique that needs a good knowledge of plant properties, chemical substances and skillful craftmanship. The importance of dyeing is obvious in the case of the Afghanistan silk workshops. The people in control of the silk trade in Afghanistan were not the silk weavers, but the dyers and, more specifically, Jewish dyers who specialised in purple and blue, the most luxurious colours (Rau 1988, 20). Different kinds of dyeing methods were employed, depending on the dyes used. The most simple one was direct, using water soluble dyes like, for example, the dye substances produced from turmeric, *Curcuma tinctoria*, safflower, *Carthamus tinctorius,* roucou, *Bixa orellana*, berberry, *Barberis vulgaris* and several lichens, *Rocella tinctoria*, etc. Sometimes lemon juice (in the case of turmeric) or urine were used in the dyeing mixture as a fixant for better results. Dyes of this category, called substantive dyes, are very fast to light, water and perspiration (Robinson 1969).

Vat dyes and mordant dyes were the ones used more often for animal and plant fibres, because they produce nicer and more varied colours. Vat dyes are "soluble in alkaline liquids and transformed into leucocompounds, which fix on the fibre by oxidation in the air without a use of a mordant" (e.g. indigo, *Indigofera tinctoria*, gives blue, *Acacia catetchu* or *Uncaria gambir* give brown, and some shellfish, *Murex trunculus, M. brandaris, Buccinium lapillus, and Helix ianthina,* give scarlet or purple). The problem with vat dyes, however, is their fastness to light and water. The most popular amongst vat dyes used to be those obtained from the shellfish *Purpura* and *Murex*, both found in the Mediterranean; a small gland of the animal contains the dye. The molluscs were broken to take out the glands, and then after a long process of salting, rinsing, boiling and skimming, the most expensive, celebrated purplish red or scarlet dye was obtained (Robinson 1969).

Mordant dyes on the other hand require help from chemical substances so as to precipitate them on the threads or the textiles. In this category fall most of the natural dyes (Robinson 1969). Red, extracted from dried insect bodies *Kermes, Kermococcus vermilia,* Cochineal, *Coccus cacti*, Lac, *Coccus lacca*, roots of plants, as madder, *Rubia tinctorum*, henna, *Lawsomia inermis*, asafoetida, *Ferula* sp., and *Alkhana tinctoria.* Different colours were produced from plants. Blue came from woad, *Isatis tinctoria*, yellow from saffron, *Crocus sativus*, ochre from *weld*, *Reseda luteola*, yellow from dyer's broom, *Genista tinctoria*, and yellow or yellow brown from sumach, *Rhus coriaria.* Several fruits have also colouring qualities: Pomegranate, *Punica granatum*, gives yellow; sunt berry, *Acacia nilotica*, and turnsole, *Croton tinctorium,* give blue; persian berries, *Rhamnus infectorius* give orange-yellow; walnuts, *Juglans regia*, give brown. Galls and insect faecal pellets from various trees give black (Robinson 1969).

Several plant dyes were used in Soufli: *Rubia tinctoria* for red, *Prunus dulcis* for yellow, *Allium sativum* L. for a carrot-like orange colour. A wide range of plants was also used by Thai village women. Dyers are described as cooks, having each a favourite way of preparing the same recipe. The ingredients they add to the dyes help to create different levels of acidity or alkalinity. They speak of their dye baths, characterising them as sweet or sour, and also use different substances as mordants as for example mud from buffalo wallows, silt, tree-bark ash or even rusted metal. Exchange of ideas and recipes took place at religious gatherings (Conway 1992, 71).

Mordants are either basic for acid dyes or acid for basic dyes. Acid mordants related to tannin are found in oak balls or bark or vegetable oils. Basic mordants are derived from salts of metals like aluminum, chromium, iron, copper, zinc or tin. Different mordants on a dye produce different results in terms of colour. Developers also were used in the dyeing process: they were chemicals applied after the dye bath in order to alter enhance or fix the colour again. Ammonia was a common chemical, and urine was widely used for dyeing purposes until recently. For example, in Thailand, after the dye baths, they use a preparation made from a plant resembling spinach: its ashes are diluted in boiling water and the mixture is strained. The silk is left to soak in it for a few days. Another recipe substitutes the above plant with ash from burnt sesame bushes, *Sesamum indicum*, after the seeds have been removed. However sesame makes the silk too shiny (Conway 1992, 79).

Unfortunately many of the natural dyes are fugitive. Some communities, however, used this disadvantage to their benefit, and they dyed the same garments with different colours for different occasions (Robinson 1969).

Weaving

Weaving consists of the following elements: a fixed set of threads, the warp, and another set, the weft. The weft is interlaced into the warp. The special frame that is used for weaving is the loom. There are lots of different types of looms, simple and more advanced. The most simple one is the back strap loom. In Egypt from around the 5th millennium BC, they used the two beam ground loom, and from there it spread to Mesopotamia and Judaea. Another popular loom was the one beam vertical warp weighted loom, used at Late Bronze Age Akrotiri and at sites around the north of the Mediterranean (Fig. 8.6). It was also used as far away as Greenland during the medieval period. The weights used for this loom were usually of clay or stone and were attached by cords to groups of warp threads (Barber 1991, 111). The larger the weights, the more threads were tied on them. Sometimes, warp weighted looms were so wide that it took two women to work on them. In this way, longer pieces of cloth were produced. The vertical two beam loom, is the most efficient and probably a late development (Barber 1991, 113).

Silk weaving sometimes involves mixing silk thread with cotton or linen or even wool: for example, in Crete during the 17-18th century AD when producing garments for the local market (Triantafyllidou-Baladie 1988); wild silk was also mixed with true silk, a process continuing in Assam, India. Weaving different patterns in silk clothes is an art by itself, and Afghanistan silks are famous for their elaborate weavings. In Soufli (Oikonomou 1992, 99), only the best pieces of clothing, had laboriously interwoven patterns with silk thread. Woollen thread was used for interweaving other secondary pieces of cloth. The cloth is sometimes named according to the patterns woven on it: "θυμιατές", censors, are named οι ποδιές με τα θυμιατά (aprons with censors). Only the filament of the cocoon is used for weaving the best clothes. Coarser pieces of silk are created with silk floss, made from the outside fibres of cocoons (Oikonomou 1992, 94, 98-99).

Weaving wild silk

Wardle (1891, 29-37), in an effort to promote the Indian wild silk, tried to prove that the difficulties in spinning and weaving the wild silk thread and the idea that it could not be compared with the mulberry silk thread, were mere prejudices. He showed that wild silk could be wound and woven easily and marketed as a secondary type of silk, only slightly inferior to Chinese material. While the mulberry silk thread is round and therefore

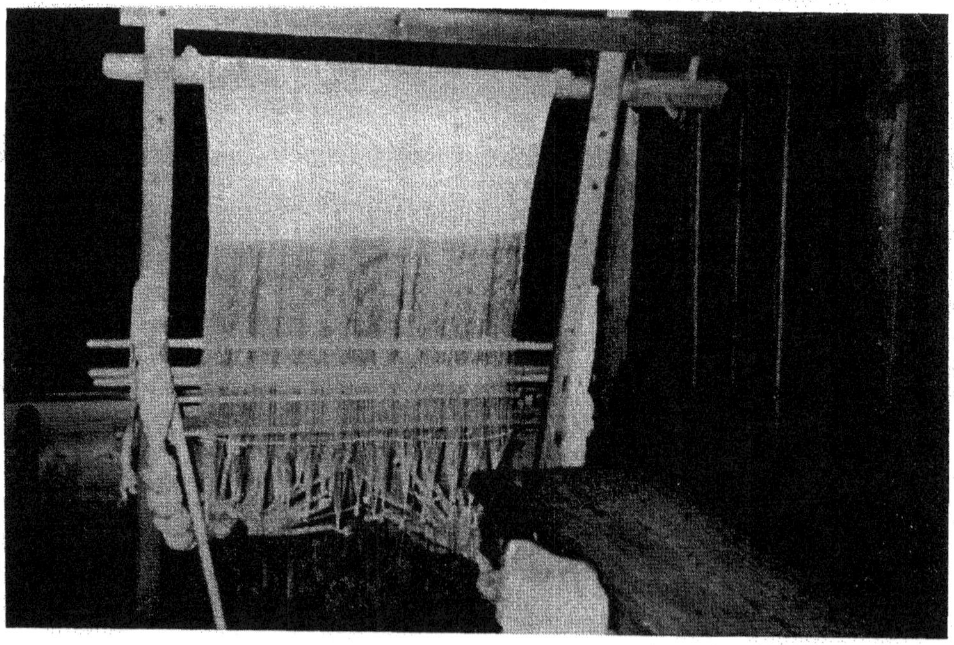

Fig. 8. 4 : Reconstruction of a warp weighted loom, Stöng, Iceland.

easier to work with, the flat thread of wild silk has a natural lustre and a shine that comes out slightly when dyed in strong colours, e.g. black or gold. Wild silk is more difficult to dye than mulberry silk, and is far more difficult to dye in a cloth piece than in the skein. As noted by Wardle (1891, 29-37), wild silk can be used for making all kinds of material, from very fine and delicate pieces of cloth, such as handkerchiefs and silk dresses and laces, to carpets and rugs and velvets. It is also used for trimming materials such as tassels and fringes.

Conclusions

Wild silk, although not as much in use today as the silk produced from *B. mori*, was probably very popular during Antiquity. Even after the introduction of the Chinese silkworm to Byzantium, Mediterranean silkworm species could have served the needs of the elite. Exploiting domesticated silkworms or gathering them in the wild, processing the thread and turning it to luxurious garment, is a whole culture. Different implements are related to processing, dyeing and making the cloth, whilst the activities associated with silk making may take the form of a cult. The evidence of wild silk in early Aegean prehistory has proved to be even more exciting.

Fig. 8. 5 : Dyeing (from Rosetti 1540).

9. Bronze Age Silk - The evidence from Akrotiri

Introduction

In the consideration of trade in prehistoric contexts, durable inorganic items, such as pottery and metal, have dominated our view of the direction, intensity and nature of exchange, mainly due to their increased probability of preservation. The scale of trade in perishable substances in the Bronze Age Aegean has only recently begun to be acknowledged (Knapp 1991), aided by the development of analytical techniques for the study of residues on the surface and within the walls of pottery vessels (Foster & Moran 1989). However, organic materials which travel without durable containers, or within organic wrappings themselves, are more elusive in the archaeological record, and interpretation of what might have formed the bulk of trade, in its broadest sense, relies upon the chance preservation of fragments in unusual circumstances. Such preservation is provided by the thick tephra deposits burying the Bronze Age settlement at Akrotiri.

Context of the cocoon

During the processing of finds from recent excavations to the east of the House of the Ladies (*cf.* Doumas 1983), from beneath the destruction levels, comes a calcified, slightly waisted, cylindrical object, 44 mm long and 18 mm wide, with flattened ends. One end is broken open revealing a smaller ovoid structure within (Fig. 9.1). It is this object which is identified as a cast of a cocoon and part of the enclosed pupa of a large lepidopteran. Unfortunately, the associated group of finds is chronologically mixed, and includes pottery ranging from the Early Cycladic period through to the phase immediately prior to the eruption. Nevertheless, the delicate nature of the object and its fine preservation implies contemporaneity with the latest material and a mid-second millennium date is probable.

Identification

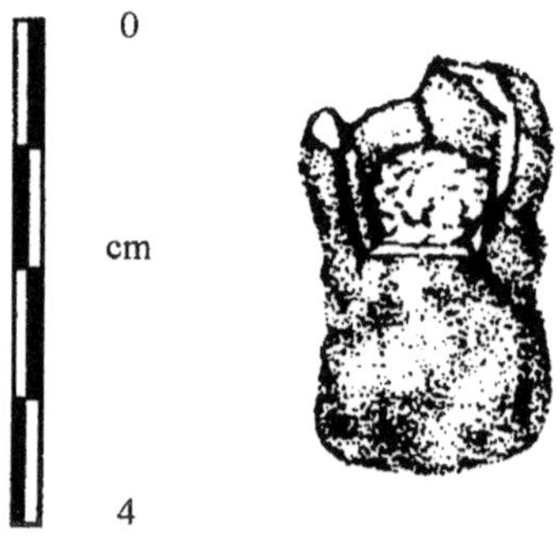

Fig. 9. 1. Cocoon and chrysalis of *Pachypasa otus* preserved by calcification, Akotiri, Thera (Drawing by A. Kontonis)

The good preservation of this find enables its identification. Several phyla of invertebrates utilise cocoons, and a number of other arthropods produce them, either to protect themselves during metamorphosis or, in the case of Arachnida, as egg cases (Roberts 1995). Among the Insecta, two Orders, Hymenoptera and Lepidoptera, most frequently construct cocoons, but they are also present in some Coleoptera and a few small Diptera. Apart from a few small skippers, the butterflies do not construct cocoons, but their use is widespread among the moths (Carter & Hargreaves 1986). In the European fauna, the size and shape of the cocoon further limits the possibilities of identification to the larger moths, which construct freely spun cocoons, in particular some species of Lasiocampidae and Saturniidae. Only a few of the Lepidoptera, however, are sufficiently large to warrant discussion in relation to the Akrotiri find and the essentially urban, or at least not rural, archaeological context further limits interpretation. The context appears an unlikely place for the natural pupation of a large moth, and it is more probable that the item was brought to the site and later discarded.

In the European fauna, two species, the saturniid, *Saturnia pyri* Den. and the lasiocampid, *Pachypasa otus* (L.), have long been suggested as the source of the 'wild silk' of the Greek and Roman world (Kirby 1903). It is probable that one of these two species produced the Akrotiri cocoon. *S. pyri* is the largest species of moth in Europe and has a largely southern distribution, from southern France, eastwards to Asia Minor and Iran (Seitz 1913). Its striking green caterpillars feed on the leaves of a range of fruit trees, including pear, *Pyrus*, apple, *Malus*, and cherry, *Prunus* species, as well as ash, *Fraxinus excelsior* L. and buckthorn, *Rhamnus catharticus* L. The caterpillar reaches a maximum length of 120mm, and is too large to have created the Akrotiri cocoon, although there is other evidence to suggest its presence. The other species, *Pachypasa otus*, which most likely formed the cocoon discussed here, is now much less common, ranging from southern Italy and Sicily to the coast of Bosnia, Greece, and eastwards through Rumania to Armenia and southwards to Israel (Seitz 1913). The caterpillars of *P. otus* feed most frequently on the leaves of cypress, *Cupressus sempervirens* L. and downy oak, *Quercus pubescens* Willd., although it has also been recorded from other oaks and juniper. The species was used for silk production in some parts of the Levant until last century (Kirby 1903).

Silk Use in the Mediterranean

The presence of pre-Han silk outside the Chinese Empire has recently been reviewed by Good (1995), and she provides an outline of the techniques used in the recovery of the silk thread from the cocoons, as well as a consideration of the problems of identifying the preserved fibre to its species of origin. The earliest known silk in the Mediterranean region is recorded from an Egyptian burial of the 21st Dynasty from Thebes dated to *ca.* 1000 BC (Lubec *et al.* 1993). However, the thread is dated by amino acids relative to the mummy and there is some doubt about the mummy's date (Ryder 1995). In Europe, silk has been claimed from a rich early Iron Age burial in the Hohmichele tumulus in Baden-Wurttemburg of the sixth century BC, and further material may come from the penecontemporaneous burials at Hochdorf-Eberdingen, near Stuttgart (Wild 1984; Good 1995). The oldest surviving material from Greece is from the Kerameikos cemetery in Athens, over a century later, and the thread from this find has been identified as imported Chinese silk from the silkworm, *Bombyx mori* L., not introduced into Europe until the sixth century a.d. (Wild 1984). Wild silk was also recovered together with the Chinese silk (Wild 1984; Good 1995). On the basis of these few finds, Barber (1991, 32) suggests that wild silk manufacture in Europe began in the Classical period in imitation of Chinese imports. Good (1995, fig. 3) claims that the modern distribution of *P. otus* includes the Aegean, but there is no clear indication that the moths were native to Santorini, rather than imported as cocoons from elsewhere, perhaps Cos or Amorgos. If Rackham's (1990) comments on the contemporary Bronze Age vegetation of Santorini are correct, there would appear to have been little possibility of intensive coppice or orchard management to provide the necessary growth for feeding the silkworms. It is possible, therefore, that the cocoon reached the island by means of trade, rather than being a local product. Ethnographic evidence in the Aegean islands includes the use of wild silk or the cocoons themselves being incorporated by embroidery into dowry items (P. Day, pers. comm.).

Pictorial evidence for the Silk Moth in the Bronze Age Aegean

One cocoon does not provide evidence for the production or exchange of silk on a grand scale, yet it may be argued that there is other strong evidence for its importance in the Late Bronze Age Aegean in both practical and symbolic terms. Indeed, it may be suggested that silk may have been a commodity whose special properties and rarity would endow both status and mystique to those able to procure it, and an association with ritual activities would not therefore be surprising.

Silk Moths and "Tree Shaking" representations

There exists a group of representations of insects which have been long referred to as butterflies (Morgan 1987, 133). These are relatively common on Late Minoan I and Mycenaean seals (Betts 1984; Younger 1983), as well as in wall paintings (Evans 1928; Doumas 1982). Nilsson (1950) discussed the 'butterflies' as an important category of symbols, while Warren (1987) has linked their occurrence on rings to the

(a)

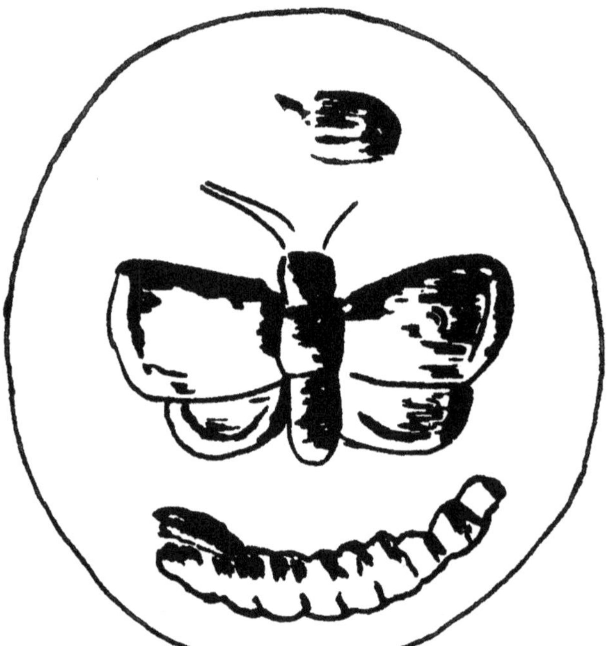

(b)

Fig. 9. 2. 'Butterflies' on gemstones : a) a Late Minoan gemstone from Knossos (Ashmolean Museum, Oxford) (after Kenna 1960). b) 1st C BC gem with imago of lepidopteran between its caterpillar (below) and ? chrysalis (Coll. Leo Menz, Switzerland) (after Davies & Kathirithamby 1989).

identification of chrysalises in the same scenes. While the interpretation of the stylised representations is difficult, the depiction of rather thick, curved antennae on a Late Minoan cut gemstone from Knossos (fig. 9.3a) is more suggestive of the hairy antennae of many moths than the filiform ones of most butterflies. Some of the distinctive wing markings on the representations also provide further features on which to base identification. In fact, such an argument can be made for most of the representations which previously have been interpreted as butterflies. On both seals (Fig. 9.4) and wall paintings, the moth is usually depicted with one or in some cases several 'eyes' on each wing. These are often also clearly bordered (Davies & Kathirithambi 1986, fig. 21; Younger 1983, fig. 47, 51; Betts 1984, fig. 187a; Evans 1930, fig. 99). Comparison with the modern specimens (fig. 8.1), suggests that the animal intended is *S. pyri.*

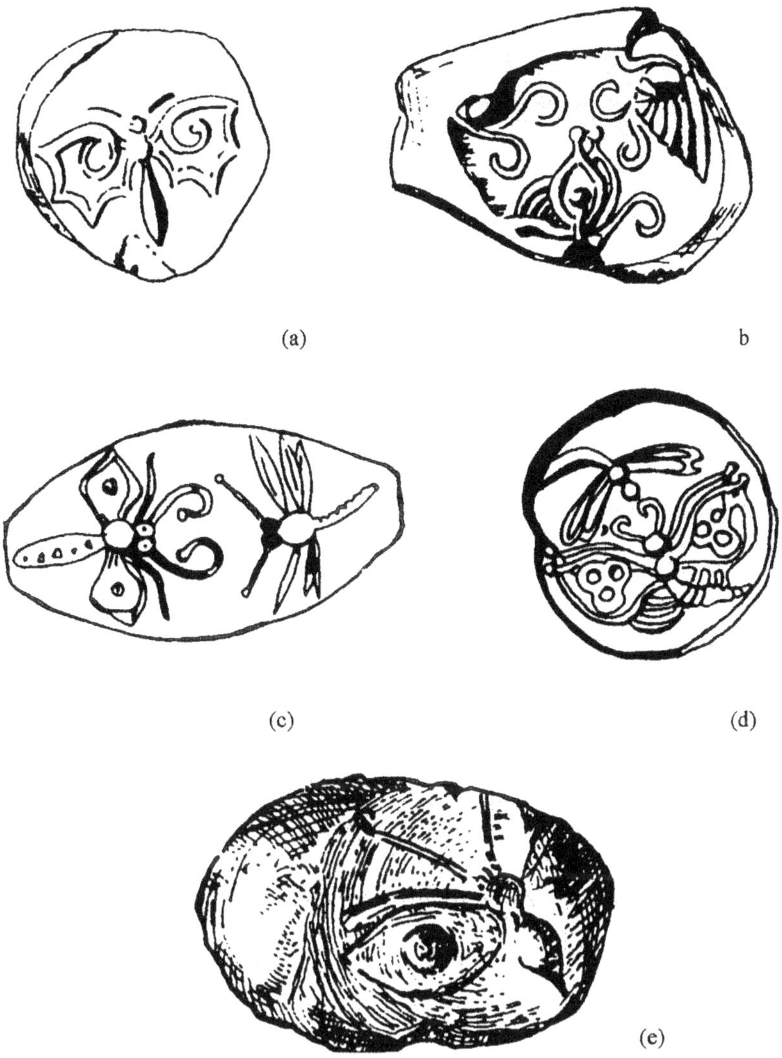

Fig. 9. 3 : Lepidoptera on Minoan seals : a & b. from Aghia Triada (after Levi 1926, in Niemeier 1992, 274). c & d. from Knossos (after Younger 1983, 128) and e. from Aghia Triada (after Evans 1930, 148).

The fact that the 'butterfly' was a popular motif during the Aegean Late Bronze Age is also confirmed by the finds of illustrations of Lepidoptera with spread wings on gold repoussé plaques from the Shaft Graves of Mycenae. Evans (1930) thought that the balance, of thin gold plate with butterflies embossed on the scales, found in the Third Shaft Grave together with golden butterflies (Fig. 9.4), was for ritual purposes, specifically the weighing of the souls of the departed. Svoronos (1906, in Nilsson 1950), in less romantic mood, however, had suggested that that they were used for weighing gold used as currency. The delicate nature of the balance implies a lighter commodity, and another possibility is that it was used to weigh the precious thread produced by the caterpillars of the moths depicted. In this case, alternative explanations are provided of the use of the moth as a symbol; through a link with souls and transformation/metamorphosis and, more simply, as a material of rarity and value. In fact, neither of these two analogies need dominate our reading of the evidence.

In later Greek myth, the use of the 'butterfly' as a metaphor for the soul is common, and in India, the wild silk moth, *Antherea mylitta,* is associated with the god Vishnou (Wardle 1891).

Marinatos argues that it is the insect life cycle, with regeneration manifested as transformation, which leads to the fascination with insects (Marinatos 1993, 195). She argues that it is for

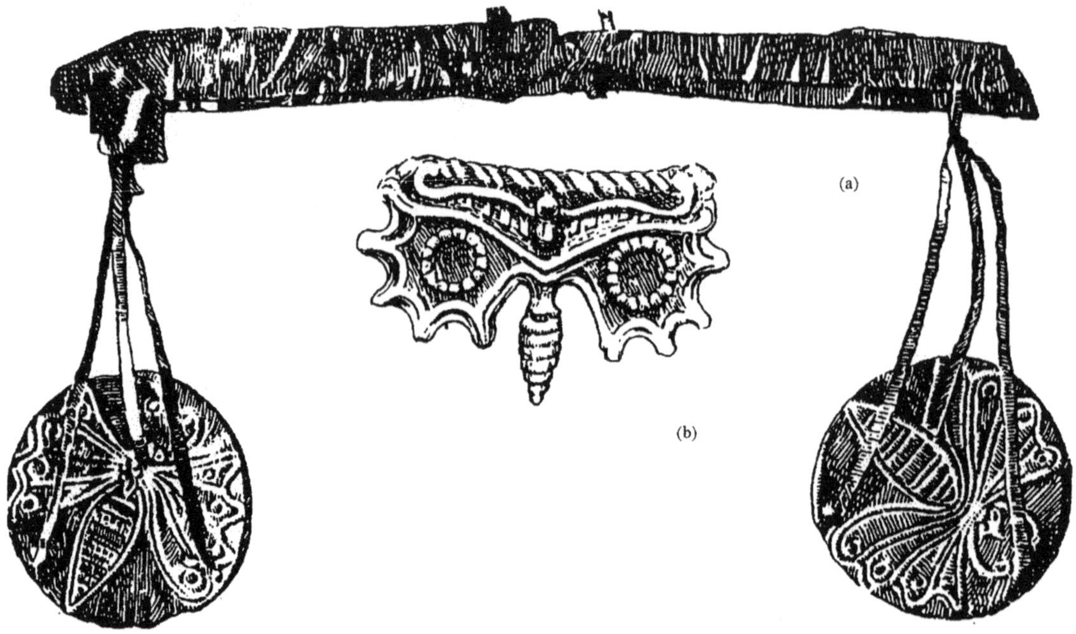

Fig. 9. 4. Balance of thin gold plate with Lepidoptera on scales (a), and (b) a gold lepidopteran pendant, both from the Third Shaft Grave, Mycenae (after Evans 1930, 159).

this reason that insect symbolism is used in representations of the tree-shaking ritual which is common in Minoan iconography, perhaps having as its focus the theme of life and death. Whatever its meanings, this link between insects and the "tree-shaking" representations is worthy of attention (fig. 9.5). Such scenes belong to a group argued to be representations of high priestesses in front of shrines (Marinatos 1993, 184-188), with the common formula of one participant bending over a pithos or stone and another pulling down the branches of a tree. Indeed tree shrines and the depictions of trees in what are clearly cult situations have led many scholars to refer to a tree cult, and to postulate links with ostensibly similar belief systems in Egypt (Marinatos 1993, 181-184; Nilsson 1950, 283). In such scenes, the insects referred to as 'butterflies' are sometimes juxtaposed with dragonflies, and frequently with shapes which have been interpreted as chrysalises (Figs 9.6 & 9.7; see also Warren 1987; 1990b, 193-206).

It is tempting to see in these repeated formulae a religious adaptation of a scene, which has its roots in the production process of silk, similar to the religious ceremonies of the Santals of India. The shaking of the tree might be associated with the collection of cocoons from the wild; the leaves of the tree on the Archanes ring and related representations are similar to oak, upon which the caterpillars of *S. pyri* and *P. otus* are known to feed. Such an interpretation would at least partly account for the repeated use of the 'butterfly' and cocoon (Warren 1990b) motifs in such scenes. A link between moth, caterpillar and pupa is provided by a later gemstone, dated to the first century BC, where all three are represented (Davies & Kathirithambi 1986, fig. 9.3.2). All together these motifs connected with the wild silk moths might suggest that the process involved in their exploitation had been expropriated somehow and imbued with a religious meaning.

Against the interpretation of "tree shaking" ceremonies (fig. 9.5) as the exploitation of wild silk might be set another consideration. It is plain that both in Aegean Bronze Age and Egyptian iconography, trees are represented as sacred objects (Nilsson 1950, 262-288). Tree shaking depicted together with images of moths and chrysalises could be related to the exploitation of wild silk.

Fig. 9. 5 : The Ring of Minos (after Evans 1935).

Silk Moth Emblems and the Ship Procession Fresco

It is instructive to move from consideration of what is usually interpreted as a religious context, to the representations of what are suggested to be silk moths in another scene; those which occur in the miniature frescos of the West House of Akrotiri: the very site where the cocoon was found.

The fresco in question shows a flotilla of six large ships with passengers and a cargo ship, which appear to be involved in some sort of festival, travelling from their departure town to a port of arrival , where people on land and in small boats wait. One of the striking features of the scene is the clear hierarchy of the ships, which is reflected in their degree of adornment. The prow and stern of the ships have been emphasised with decorative elements which might be seen as motifs for interpretation by the viewer of the wall painting, or as emblems added to the boats for the occasion of the nautical procession (Morgan 1987; 131). All of the ships carry a 'star emblem' (see below), but they also all carry animal emblems; such as dolphins, birds and what have been identified as butterflies. All of these have been viewed as images of swift movement through sea and air and therefore as being appropriate metaphors for transport (Morgan 1987; 133).

Fig. 9. 6. Lepidopteran and chrysalis on the Vapheio Ring (detail after Evans 1930, 148).

Fig. 9. 7 : Lepidoptera and chrysalises on the Ring of Nestor (after Evans 1930, 148).

Among the most prominent motifs are those suggested here to be moths. These are applied to Ships 1 and 2. Ship 1 has one on the prow, whilst Ship 2 has two on the prow and two more on the mast. The moths on the two ships are of different types, one with wings with a scalloped edge and the other with pointed wings with a smooth contour (fig. 9.8, 9.9). Ship 2 is clearly the principal vessel, the 'flagship' and so the fact that it has four of the moth emblems should allow these to be seen as an indication of prestige. The individuals on the flagship are connected with a large number of symbols of power, such as the boars tusk helmets and the Egyptian 'waz' motif (Morgan 1987, 23). Perhaps again the wild silk moth is seen here as being connected with the exotic or special, as an elite symbol. Another symbol from the Ship Procession Fresco, that of the star, will be reassessed later, but first further evidence for silk in the Bronze Age Aegean will be considered.

Pictorial evidence for silk cloth in the Aegean

Women depicted in Minoan wall paintings often wear a diaphanous blouse under an open bolero (Ventris & Chadwick 1973). It is possible that these diaphanous clothes, including those worn by the ladies in the Akrotiri wall paintings, were made of wild silk. Silk thread is finer than the finest linen and is more likely to produce transparent, light, high quality clothes. The appearance of transparent garments on seals, rings and wall paintings could imply the extensive use of silk in the clothing of the ruling classes in the Aegean Late Bronze Age.

The star symbol on the ships : evidence for another textile?

The moths, previously referred to as butterflies, on the Ship Procession Fresco have formed part of the reinterpretation of the role of silk in the Bronze Age Aegean. Alongside these symbols on Ships 1 and 2, occur a star or rosette emblem (Fig. 9.8, 9.9), which is attached to the bowsprits, not only of those ships, but of each vessel in the procession. Morgan (1987, 132), in a full discussion of the star motif, relates it to star designs on LMIB Marine Style and other pottery, continuing the nautical connection. If the moths on the prows of the ships are seen as silk moths, then other elements in the iconography should be re-examined. Morgan has interpreted the objects on the rigging as pendants of "stylized crocuses" (1987, 29), but suggests that the feature associated with the 'butterfly' on the prow is a 'star', making it virtually the only motif used to decorate the ships which represents an abstract feature. Diapoulis (1980) identified these objects as flowers of the lily *Pancratium maritimum*, although it bears no resemblance to the lilies in the other wall paintings (e.g. the Spring Fresco). It is suggested that the object appears far more similar to the seed head of cotton, *Gossypium sp.*, an interpretation more in keeping with the other elements in the Ship Procession Fresco, which imply a link to textiles.

Fig. 9. 8 : 'Butterflies' and the ? cotton motif on the flotilla fresco, Akrotiri, Thera.

On the history of Cotton

The history of cotton in the Mediterranean is as uncertain as that of silk. Zisis (1955) reports cotton from the site of Trakhones in Attica during the fifth century BC, and Barber (1991, 33) suggests that it did not reach the region until early in the Classical period, whilst Arkell (1971) ascribes its arrival to Meroitic Egypt; both support an Asiatic origin for the crop. Tree-cotton, *Gossypium arboreum*, is held sacred by Hindus, and it is quoted in the laws of Maru that may go back to 800 BC (Forbes 1956). It appears to have been cultivated by *ca*. 5000 BC at Mehrgarh on the edge of the Indus plain, and woven fabric has been recovered from Mohenjo-Daro, in the area of the Harappan civilisation at least as early as 1500 BC (Allchin & Allchin 1982). 'Trees bearing wool' are referred to in an inscription of Sennacerib of 694 BC at Nineveh, and these seem likely to have been tree cotton (King 1909). The Classical sources (Theophrastus, *His*. 4,4,8; Herodotus, 3, 106; Pliny, *HN*, XII.22.39) would also support an origin in India (*cf.* Potts 1992, 133-134) On genetic grounds, however, Wrigley (1960) has speculated that cotton was cultivated in Egypt before 3000 BC, and there is some support for this from the archaeological record. In Upper Egypt, Chowdhury and Buth (1971) have identified the seed of native cotton, *Gossypium herbaceum*, from dung dated to *ca*. 2500 BC, and a Sudanese origin for the crop is favoured by Harlan (1992, 72). Whilst the Nubian cotton was evidently being used as cattle fodder (Chowdhury & Buth 1971), it is probable that cultivation for its fibres in the Nile valley is of some antiquity. Betts *et al.* (1994) have recently published evidence of cotton fabric from Jordan with a date of 4450-3000 cal. BC. and it may have been more widely known in Antiquity than sources and the slight fossil record would suggest.

Whether the scene in the ship procession represents the spring arrival of a trading fleet or as seems more likely, a celebratory regatta is impossible to know, but the African component of many other Theran and Minoan frescoes may be relevant to the extent of trade and contact with Africa (Vanschoonwinkel 1990).

Spindles and spindle whorls

In the context of industry on Santorini, the virtual absence of spindle whorls has been thought problematic (Tzachili 1992); in the West House, only four were found. Exactly the same phenomenon has been noted in Middle and Late Minoan Crete (Carington-Smith 1975), although at Phylakopi ten whorls and seventeen 'weights' have been recovered. However, implements for textile making, which have so far been discussed concern flax and wool fibre manufacture. Cotton is spun with different types of spindles than those for bast fibres and wool, while silk is often unspun, merely being unwound from the cocoon. Also the absence of spindle whorls could be as easily attributed to the perishable material from which they were made, as to the use of a different technique for spinning. The extent of excavation also cannot preclude the possibility that the spinning of thread may have taken place elsewhere. As cotton fibres are short and fine, spinning for cotton is characterised by the use of a very light supported spindle weighing an ounce or less. With the use of such spindles the Indian spinners were able to make more than 200 miles of thread from a pound of cotton (Forbes 1956, 156; Barber 1983, 43). The spindle whorls used for cotton are also very small; the smallest ones recorded are Islamic ones, of a diameter around 8mm and weight less than a gram (Liu 1978, 90-91). In an archaeological context, they could easily be mistaken for beads.

Fig. 9. 10. 'Butterflies' with scalloped wing edges, Flotilla Fresco, Akrotiri, Thera.

Conclusion

The identification of the silk moth cocoon, archaeoentomological research at Late Bronze Age Akrotiri has revealed the presence of wild silk in the Aegean earlier than the mid second millennium BC, and its use in a variety of contexts has been suggested. The cocoon, like the range of stored product insect pests recorded from the site, is evidence of the presence of high value commodities probably traded into the island which might otherwise pass unnoticed in the archaeological record. There is, however, no need to see Thera's involvement as direct. On a broader level, the reinterpretation of the 'butterfly' motif in the Aegean is of importance. The chrysalis features on some of the same scenes, perhaps along with "tree shaking", and seems to link the process of silk exploitation to religious ceremony. The moth motif and the putative cotton flower are clearly used in the flotilla wall painting as symbols of power and prestige; the silk moth is also a repeated iconographic representation elsewhere.

Through a link in the iconography of the Thera frescos, and specifically the identification of the 'star' emblem on the Ship Procession Fresco as cotton, it has been suggested that cotton may have been of importance, perhaps as an imported material from Egypt. Such a connection would not be out of place with other evidence of African and/or western Asiatic connections for Akrotiri with such as the rhyta made from ostrich eggs (Sakellarakis 1990).

Thera itself has plentiful evidence for its involvement in textile manufacture and trade. If the implication of the moths and cotton flowers in the wall paintings is one of trade in raw materials, Akrotiri might be seen as involved in the production and export of the finished textiles. In the limited area of the settlement excavated, there are indications of some specialisation in textile manufacture. One house has produced 450 weights for upright looms (Tzachili 1990), and the occurrence of the shells of *Murex brandaris L.* and *M. trunculus L.* (Karali-Yiannakopoulou 1990) probably reflects the production of scarlet and purple fabric dyes, whilst yellow is implied by the wall painting of the crocus gatherers (Douskos 1980). There has been much discussion of the status of the settlement in Thera (e.g. Doumas 1983), but its involvement in the trade of a variety of commodities is generally accepted (Hardy *et al.* 1990).

10. Bees and beekeeping

Introduction

Honey and beeswax were, and still are, important trade commodities, and beekeeping must have been of great significance for the economy of early communities. Whilst in India, cane sugar has been used since early prehistory (Brothwell 1969, 83-84), it appears only to have become known to the Western World in the second and third centuries AD (*idem*). The Venetians were the earliest refiners of sugar cane in Europe, from the fifteenth century; and the technique of refining sugar from beet was only discovered in 1747 (*idem*). Honey was the usual sweetening material and its great importance in Antiquity is attested in Classical texts. Substances which acted as sugar substitutes are recorded in Herodotus, and even sugar is noted in Pliny as different kinds of honey. This review is based largely upon Crane (1983), as a prelude to the evidence from Akrotiri. Doumas (1983, Fig. 10.1) described a hive from Late Bronze Age Akrotiri, and here is presented evidence of a further type of hive, modelled in miniature.

Fig. 10. 1. Pottery vessel interpreted as a beehive from Room Delta 17, Akrotiri, Thera (scale 1:8) (after Doumas 1983).

Bees and honey

Honey producing bee species

Apis mellifera L. is the most common honey bee species, widely distributed and indigenous to Europe, western Asia and Africa. Other honey producing bees include *A. cerana*, very much resembling *A. mellifera,* but smaller and, as a result, building smaller nests and produces less honey; it is native to Asia. *A. dorsata,* on the other hand, is a large bee that builds a single comb in the open, hanging from the branch of a tree or the edge of a rock. It needs light and cannot live in a dark hive. *A. florea*, makes a single comb nest and lives also in the open. Another group of bees are the *Meliponini*. In contrast with the species already examined, the *Meliponini* bees are stingless and native to America, Australia, Africa and Asia (Crane 1983, 13-18).

Bees live in clusters, the size of which varies throughout the year. What determines the size of a hive is firstly, the number of the bees, and secondly the need to store honey for the dearth season. The insects spread out more when the nest is hot so as to prevent overheating and contract into a small space when it is cold; during very cold winters the bees occupy one tenth of the space they normally use. Honeybees create new colonies by swarming, but the colony will then produce only half as much honey for the beekeeper than if he had prevented swarming. Modern beekeeping management aims at high populations without swarming.

First evidence for the use of honey

Mesolithic rock paintings at El Montero, Tervel Province, Spain dated to *ca.* 8.000-5.000 BC show the collection of wild bee honey, and a similarly dated painting, showing *A. dorsata,* is known from Bhimberkah in Central India, where there are also later illustrations (Crane 1983;19-25).

Some of the earliest representations of bees

At the site of Çatal Húјuk in Anatolia paintings of bees occur in a Neolithic context of *ca.* 7000 BC (Melaart 1962). However they provide no information about beekeeping or honey collecting, as they simply depict the insects. Around the Mediterranean basin, the bee symbol became a cult figure, and Neolithic representations of a bee goddess occur at Neolithic Proto-Sesklo (Greece) and Starçevo (Hungary). In Egypt, a bee was used as the symbol of the Lower Kingdom (Crane 1983, 43), and Gimbutas (1974) claims that the bee goddess is inscribed schematically on the head of a bull made of bone from north-west Ukraine which has been dated to 4000-3000 BC. Greek Minoan and Mycenean art has many representations of bees. One of the most significant is the gold pendant from Mallia, dated to 2000-1700 BC (Hood 1976) showing two bees or wasps face to face (Fig. 10.2). Discussion has taken place over the identification of the insects. In spite of the arguments presented by Richards (1974; 1975) for identifying the animals as wasps, of the genus *Polistes,* it is more likely that they are honey bees, and the basic form of the representation is close to that of the Egyptian hieroglyph '*bit*', the bee (*cf.* Wilkinson 1992, 114-5).

Fig. 10. 2. Gold pendant from Mallia, Crete, dated to *ca.* 2000-1700 BC with two Hymenoptera, ?bees (redrawn from a photograph, original in the Archaeological Museum, Heraklion, Crete).

Honey as food and as a preservative

The earliest archaeological evidence for honey from Europe, comes from a residue in a Neolithic pot from Runnymede, on the river Thames, west of London (Needham & Evans 1987). The earliest European written references to honey appear on the Linear B tablets (Ventris & Chadwick 1973). It is spelt me-ri-to (Pylos 171=Un 718) or short me-ri (the equivalent modern Greek word is μέλι). It is one of the commodities weighed in liquid measures, and appears very often in lists of edible commodities. Also, there are records of jars of honey given as offering to the deities, for example that on tablet Gg702 from Knossos (Ventris & Chadwick 1973, 308-310):

pa-si-te-o-i / me-ri AMPHORA I
da-pu?-ri-to-jo/po-ti-ni-ja "me-ri"
AMPHORA I

One jar of honey to all the gods,
One jar of honey to the Mistress of the
?Labyrinth

Honey and ambrosia were regarded as the food of the gods in Ancient Greece. As honey is perishable, it is rarely recorded in the archaeological record. Most of our information comes from Egyptian tombs, where the conditions of preservation are optimal (Zander 1949). As well as the Neolithic Runneymede find (Needham & Evans 1987), there is also a Late Bronze Age record from a barrow at Ashgrove, in Scotland dated to *ca.* 1000 BC (Zander 1949). The more widespread use of gas chromatography and other techniques in the analysis of pot contents is likely to provide more information about the extent of use of honey. It is as likely to have been used as a sealant for porous clay vessels as food. In addition, it was used as a basis for alcoholic beverages. Small honey cakes shaped into figures, made of flour, honey and spices, dated to *ca.* 1400 BC, have been found at Dier el Medinah, in Egypt (Crane 1983, 239). At Akrotiri on Santorini, they still make "melitinia", round cakes of flour, honey and spices, every Easter.

Honey was also used in Egypt for preserving the dead. A story about the use of honey as a preservative is noted by the Arab writer Abd al-Latif (in Wallis Budge 1925, 208). Thieves in the Pyramids, came across a sealed jar. They opened it and when they found it full if honey they started eating it. "Someone in the party remarked that a hair in the honey was turned round one of the fingers of the man who was dipping his bread in it, and, as they drew it out, the body of a small child appeared with all its limbs complete, and in a good state of preservation; it was well dressed, and had upon it numerous ornaments." In the Classical world, the use of honey for the preservation of dead bodies was also known. Agesilaus, King of Sparta, after dying in Libya in 360 BC, was brought back to his home town to be buried; his corpse was preserved in honey. Alexander the Great's body was also placed in a golden coffin, in honey (Strabo, in Crane 1983, 240). Crane (1983, 240) also refers to the use of what was seen as a semitic custom at St. Peter's church, Titchfield, Hampshire, where the bodies of the first four Earls of Southampton (the first died in 1550 and the fourth in 1667) were put in lead coffins full of honey.

Beeswax

Beeswax is produced by worker honeybees from wax glands on the underside of the abdomen. Keros, beeswax, is recorded in the Linear B Pylos Ma- tablets and in the Knossos Mc- tablets. It appears to be measured by weight units (Ventris & Chadwick 1973, 290, 302, 465). In Antiquity, it was used for a number of different purposes. The "magical" beeswax figures from Ancient Egypt, now in the British Museum and the Agricultural Museum of Cairo, and dated to *ca.* 1400 BC, are some of the earliest surviving attempts at wax modelling (Crane 1983, 242). Beeswax was much used for votive offerings, as thanksgivings for recovery from an illness or accident in German folk art. A collection of such offerings survives in the Heimatmuseum at Rosenheim, south east of Munich (Hansmann & Hansmann 1959).

Beeswax was well known at least from Roman times onwards for making seals. Sometimes an additive was mixed with the beeswax so as to harden it, and verdigris or vermilion could be used to colour it. It also served as an adhesive. In the myth of Daedalus, he and his son, Icarus managed to fly, attaching wings of feathers to their bodies with wax. Beeswax was the material used to bind the charcoal black pigment on the head of Nefertiti (Egyptian Museum, East Berlin). Encaustic painting, wax applied into pigment, is an ancient Egyptian technique which remained popular into the Roman period (Anon. 1978). Batic, is the method of printing coloured designs on textiles by covering the areas not to be dyed with wax, before applying the dye to the fabric. It is a craft based on beeswax, and is still employed in order to produce patterns on textiles (Gillow & Barnard 1991).

Beeswax is also used for producing moulds for metals, by a technique called cire-perdue. The object to be cast in metal is first made in beeswax; this is coated with a refractory clay. Afterwards, it is heated until the wax has melted and run out; the liquid metal is poured into the cavity. The *cire perdue* technique was used by the Sumerians and the Egyptians before 1000 BC, and also by the Greeks, who developed variations of wax casting (Mills & Gillespie 1951; Crane 1983, 245). By the Iron Age, the technique was widely employed in the production of copper alloy artifacts in Europe (*cf.* Foster 1995). Statues and miniatures, artifacts and weights were made with the use of this technique. For example, in West Africa it was used for casting brass by the Ashanti tribe. The brass weights, known as the "Ashanti gold weights" because they were used for weighing gold dust, had the form of tiny figures related to situations of everyday life (Crane 1983, 246).

Early evidence for beekeeping

Egyptian hives: ancient and modern

The earliest evidence for beekeeping comes from Egypt. Four very similar scenes dated from *ca.* 2400 BC, 1450 BC (two), and 600 BC have been recovered. The earliest is a stone bas relief in the Sun-temple of Niuserre at Abu Gurab (5th dynasty, Old Kingdom), which shows a villager kneeling in front of a row of hives (Strouhal 1996, 115). Later examples, in tombs on the West Bank, opposite Luxor, 600 miles up the Nile show that horizontal pottery hives were used in ancient Egypt by the New Kingdom (*idem*); the size ratio of these hives was 2.9 x 3.7 or 1.5 x 3.7 (Crane 1983). Egyptian hives were very practical as they had two openings and combs could be removed easily by cutting them either from one end or the other. The Egyptian hives are the first step towards the movable comb hives, and most probably are the reason for the extent of knowledge of beekeeping during Classical times. Pottery dishes, one inserted over the other and sealed together for keeping the honey were also used. Not much has changed over the last 4,000 years. Today mud hives are still in use, their usual size is 137 x 17 cm, and they are placed one over the other in an effort to avoid overheating and save space. They are difficult to move and break easily but, on the other hand, they are cheap, and easy to make and manage.

Anatolian records

The Hittite laws, probably first drafted *ca.* 1650 BC, include a paragraph on the theft of bees (Saggs 1989, 166). Also from eighth century Assyria there is a record of the introduction of beekeeping. An inscription, set up by the governor of Euphrates, has the following:

> "I am Shamash-resh-usur, Governor of Suhu and Mari. Bees which gather honey which no one among my predecessors had seen nor brought down to the land of Suhu, I brought down from the mountains of [eastern Turkey] and established in [my town]. They collect honey and wax, and the gatherers also understand it. Let any person who comes along subsequently ask the old men of the country whether it is true that Shamash-resh-usur the Governor of Suhu introduced bees"

(in Saggs 1989, 216).

Greek hives

Among other professions the one of beekeeper is mentioned in the Linear B tablets (Ea820, Ventris & Chadwick 1973, 134). Doumas (1983) has described a pottery hive retrieved from Akrotiri (fig. 10.1), and other examples of hives come from mainland Greece (Graham 1975) Residues from pots interpreted as hives, from Vari, near Athens, were tested for beeswax with positive results (Graham 1978). Several were recovered from the Athenian Agora (400 BC). Two complete hives were used as a coffin for a child at Marathon (Jones 1976), *ca.* 200-100 BC. Another example was found in the Justinian fortress at Corinth (500-600 AD). The Greek hives look like the Egyptian ones in shape, only they are wider at one end. They are made of terracotta, lighter and easier to move, as well as more durable. These kinds of hives were in use until recently in the Cyclades. Another type of hive used in Crete is the 'cannon hive', a long horizontal cylinder open at both ends, and closed with pine disks. It should also be mentioned that for the inside of the hive, a movable comb technique was used. The same technique was used in Egypt (Crane 1983, 39). Compared to the Egyptian hives, the Greek ones seem like a development of the same invention, wider at one end. Hive extensions were used to increase their capacity.

Classical writers on hives

Among the classical writers, the first to give extensive descriptions and information about bees is Aristotle. There are many references on bees in Aristotle's work; the most extensive ones are in *Historia Animalium* (5. 21-23 and 9. 40) and *De Generatione* (3.10). Although Byl (1980, 16-25, 27, 48, 59, 61, 63) criticises Aristotle for being influenced by popular beliefs and for not giving a proper scientific account, Aristotle's knowledge on bees is admirable. The Arab writer Al Damiri Kamal al Din (*ob.* AD 1405), in his encyclopedia of animals, preserves a tradition that has Aristotle watching the bees through a transparent, possibly horn hive. Fraser (1931) believed also that Aristotle made his observations using a top-bar hive and the Greek bee-keeper N. J. Nikolaidis (1959) shared the same opinion.

Fig. 10. 3. Beehives and bees on a coin from Dyrrachium Obrimi (from Claudio Menetreio (1657) *Symbolica Dianae Ephesiae*, after Crane 1983).

A variety of Roman hives is described by Classical writers. Most were horizontal. Columella and Palladius discussed log hives. Hives made of cork bark are also referred to by Pliny, Varro, Columella, Palladius, and Virgil; Pseudo Quintilian mentions woven wicker hives. Other types mentioned include dung hives (Columella). Earthenware hives were not favoured by Columella because they were too hot for the bees during summer and too cold in winter and brick hives were criticized by him for not being movable. Pliny describes a type of transparent hive made either from mica or a kind of horn used for lamps. Other hives made of wooden boards (Columella, Palladius), and the ones made of fennel (Varro, Pliny) could have been vertical. Hives, according to Virgil (Georg. IV 33-36), were made of cork bark sewn together, or of woven osier, like a straw skep.

Aristotle and Varro, as well as Nicander's now lost work Μελισσουργικά (on bee keeping), must have been Virgil's background literature for his reference to bees in the Georgics (IV 315-424). The passage on bees is named "bougonia", born by a bovine, and describes how Aristaeus discovered a method of getting a swarm of bees from a bull carcass.

All ancient writers, except Aristotle, agree that bees could be born from a dead ox treated in a certain way. However bees very rarely visit carrion, and only in places where vegetation is extremely scarce would bees nest in dried carcasses, as the dried carcass offers a similar environment to the one of a hollow tree for example. This has been recorded in Egypt and other parts of Africa, particularly in places where there are no trees. It is more probable that ancient writers had confused *Apis mellifera* with a Syrphid fly, the hornfly, *Eristalis tenax* L., that lays its eggs on carcasses. The fact that this belief carried on until the seventeenth century, shows how strong it was among common people.

Pliny, Aelian and Columella also wrote upon bee-keeping:

> *"Alvaria orientem aequinoctialem spectare convenit. aquilonem evitent,*

nec favonium minus. alvus optima e cortice, secunda ferula, tertia vimine; multi et speculari lapide fecere, ut operantes intus spectarent. circumlini alvos fimo bubulo utilissum, operculum a tergo esse ambulatorium, ut proferatur intus, si magna sit alvus aut sterilis operatio, ne desperatione curam abiciant, id paulatim reduci fallente operis incremento. alvos hieme stramento operiri, crebro suffiri, maxime fimo bubulo."

Pliny (HN XI XLVII. 80)

"It is well for the apiaries to look due east, and to avoid the north wind as well as the west wind. The best hive is made of bark; the next best material is giant fennel, and the third is osier. Many too have made hives on transparent stone, so that they might look on the bees working inside. It is very useful for the hives to be daubed all over with cow dung, and for a movable cover to be made of dung, that it may be brought forward if the hive be large or the working unproductive, lest the bees lose hope to ease care; this cover should be gradually slid back so that they do not see how their work has grown. In winter cover the hives with straw, and fumigate them repeatedly, especially with cow dung."

(trans. by Rackham 1940)

Columella has a chapter on bees in his book about agriculture. The choice of beehives is covered extensively in book IX VI :

"Igitur ordinatis sedibus, alvearia fabricanda sunt pro conditione regionis. Sive illa ferax est suberis, haud dubitanter utilissimas alvos faciemus ex corticibus, quia nec hieme frigent, nec candent aestate; sive ferulis exuberat, iis quoque, quod sunt naturae corticis similes, aeque commode vasa texuntur. Si neutrum aderit, opere textorio salicibus connectentur: vel si nec haec suppetent. lingo cavae arboris aut in tabulas desectae fabricabuntur. Deterrima est conditio fictilium, quae et accenduntur aestatis vaporibus, et gaelantur hiemis frigoribus. Reliqua sunt alvorum genera duo, ut vel ex fimo fingantur, vel lateribus extruantur: quorum alterum iure damnavit Celsus, quoniam maxime est ignibus obnoxium; alterum probavit, quamvis incommodum eius praecipuum non dissimulaverit, quod, si res postulet, transferri non possit. Itaque non assentior ei, qui putat nihilo minus eius generis habendas esse alvos: neque enim solum id repugnat rationibus domini, quod immobiles sint, cum vendere aut alios agros instruere velit; (hoc enim commondum pertinet ad utilitatem solius patris-familias) sed, quod ipsarum apium causa fieri debet, cum aut morbo aut sterilitate et penuria locorum vexatas conveniet in aliam regionem miti, nec propter praedictam cuasam moveri poterunt, hoc maxime vitandum est."

"When, therefore, the sites have been arranged, beehives must be constructed in accordance with local conditions. If the place is rich in cork-trees, we shall certainly make the most serviceable hives from their bark, because they are neither cold in winter nor hot in summer; or if it grows plenty of fennel stalks, with these too, since they resemble the nature of bark, receptacles can be quite as conveniently made by weaving them together. If neither of these materials is at hand, the hives can be made by plaiting withies together; or, if these are not available either, they will have to be made with wood of a tree either hollow or cut up into boards. Those made of earthenware have the worst qualities of all, since they are burnt by the heat of summer and frozen by the cold of winter. Two kinds of hives remain to be described, those which are either made of dung or built of bricks. Celsus was right in condemning the former because it is very liable to catch fire; the latter he approved, although he made no secret of its chief disadvantage, namely, that if occasion should arise, it cannot be moved to another site. I do not agree with him who thinks that hives of this kind ought to be used in spite of this drawback, for it is not only against the interests of the owner that they should be immovable when he wants to sell them or furnish another site with hives (for these considerations concern the convenience of the owner alone), but the question arises as to what ought to be done for the sake of the bees themselves, when it is advisable that they should be sent to another district because they are suffering from disease or from the barrenness and poverty of the locality and yet cannot be moved for the reason mentioned above-a state of affairs which ought above all things to be avoided."

(trans. by Forster & Heffner 1954)

Finally Varro (RRm III. XVI.15) gives the following information about hives :

"Ubi sint, alii faciunt ex viminibus rutundas, alii e ligno ac corticibus, alii ex arbore cava, alii fictiles, alii etiam ex ferulis quadratas longas pedes circiter ternos, latas pedem, sed ita, ubi parum sunt quae compleant, ut eas conangustent, in vasto loco inani ne despondeant animum. Haec omnia vocant a mellis alimonio alvos, quas ideo videntur medias facere angustissimas, ut figuram imitentur earum. Vitiles fimo bubulo oblinunt intus et extra, ne asperitate absterreantur, easque alvos ita collocant in mutulis parietis, ut ne agitentur neve inter se contingant, cum in ordinem sint positae. Sic intervallo interposito alterum et tertium ordinem infra faciunt et aiunt potius hinc demioportere, quam addi quartum. Media alvo, qua introeant apes, faciunt foramina parva dextra e sinistra. Ad extremam, qua mellarii favum eximere possint, opercula imponunt. Alvi optimae fiunt corticeae, deterrimae fictiles, quod et frigore hieme et aestate calore vehementissime haec commoventur."

"Some build round hives of withies for the bees to stay in, others of wood and bark, others of a hollow tree, others build of earthenware, and still others fashion them of fennel stalks, building them square, about three feet long and one foot deep, but making them narrower when there are not enough bees to fill them, so that they will not lose heart in a large empty space. All such hives are called alvi, "bellies", because of the nourishment, honey, which they contain; and it seems that the reason they are made with a very narrow middle is that they may imitate the shape of the bees. Those that are made of withies are smeared, inside and out, with cow dung, so that the bees may not be driven off by any roughness; and these hives are so placed on brackets attached to the walls that they will not be shaken nor touch one another when they are arranged in arrow. In this method, a second and a third row are placed below it at an interval, and it is said that it is better to reduce the number than to add a fourth. At the middle of the hive small openings are made on the right and left, by which the bees may enter; and on the back, covers are placed through which the keepers can remove the comb. The best hives are those made of bark, and the worst those made of earthenware, because the latter are most severely affected by cold in winter and by heat in the summer."

(trans. by Hooper 1935).

On bee-keeping

Extracting honey out of hives

In very early times, beekeepers were already able to remove the honey from a horizontal hive, a few combs at a time from either end of the hive. In traditional upright hives, honey was stored above the brood nest, between it and the attachment of the top of the combs to the hive itself, so it was difficult for the beekeeper to extract it. A door in the side of the tree or the use of a cap above a skep were ways that facilitated the removal of honey. The use of a door in log hives is a necessity, and has been used since very early times in Russia, Poland, Germany (Crane 1983, 97). Wicker skeps are a very common type of beehive mainly used in Western Europe, and the remains of one are known from the Iron Age site at Feddersen-Wierde in northern Germany (Ruttner 1979). The widely reported example from Anglo-Scandinavian York has been doubted by Kenward (1991; Kenward & Hall 1995, 765-767), despite abundant fossils of *Apis mellifera* in the deposits. Honey could be harvested from a skep without killing the bees by driving. This would be done in the autumn, when there was little brood in the hive, since it is very difficult to induce bees to leave any brood they are covering. The skep to be harvested was inverted, and an empty skep was placed above it, in contact at one point and elsewhere held away from it at an angle. A skewer and a pair of diving irons were used to fix the skeps in position. Driving could also be used to move some of the bees (into the upper skep) to start a new colony. The queen was looked for, and the skep without her would be given a queen in a cage (Crane 1983).

Moveable comb hives: an "innovation"

As already described, the bees build the honey combs downwards from the top of the interior surface of the hive and attach them to the sides as well. The beekeeper can remove a comb only by cutting it out. This is difficult when dealing with horizontal hives, but it is even harder with upright hives where the access was at the bottom, well away from where the comb was attached. The movable comb hive was the solution to the big problem of honey extraction. It is vertical, and its advantage compared to other vertical hives is a series of wooden bars put on its mouth, where the honey combs are attached. Sir George Wheler first described the use of movable comb hives in his book "A journey into Greece" in 1682:

> "The Hives they keep their Bees in are made of Willows, or Osiers, fashioned like our common Dust Baskets, wide at the Top, and narrow at the Bottom, and plaister'd with Clay, or Loam, within and without. They are set the wide end upwards, as you see here. The Tops being covered with broad flat Sticks, (as at C.C.C.) are also plaistered with Clay on the Top; and to secure them from the Weather they cover them with a Tuft of Straw as we do. Along each of those sticks the Bees fasten their combs: so that a combe may be taken out whole, without the least bruising..."
>
> (Wheeler 1682)

The biggest asset of the movable comb hive is that honey combs can be removed fairly easily. Also it can be inspected, cleaned, and during springtime, the colony of the bees can easily be divided into two parts. Hives of this form are still used in Crete, and are either made of terracotta or wicker.

Top bar hives are not mentioned by Classical writers, while the difficulties in feeding the bees and cleaning the hives are acknowledged as a problem. Aristotle's account implies the existence of movable comb hives. Ifantidis (1983) has suggested that the movable comb hive might have been preceded by a movable nest hive. For example, if the lid is placed over the wide mouth of a hive, when removed, the combs will come out with it. A whole range of innovations, after the widespread adoption of the Greek top bar hive lead to the movable frame hive.

The miniature 'bee-hive' from Akrotiri

Two miniature vessels (Fig. 10. 4) were recovered in Xeste 3, Akrotiri. The vessels are roughly 15 cm high and 10 cm wide. They have cylindrical bodies and their upper parts are conical. Ovoid slats are modelled on these tholos-like pots, suggesting a kind of roof made of wooden shingles. A round opening is found on the roof. Marinatos (1976, 31) notes that the vessel recovered was wheel made, of a yellowish clay that does not exist in the island, and consequently they are not of Theran origin. He thought that the vessels were either granaries, or more probably model beehives. The granary argument is problematic. Granaries of this type have a small opening at the top and a larger, door-like opening door near the bottom (e.g. fig. 5. 3 above). The shape of the vessel and the place the opening is situated makes it more likely that it represents a beehive, and it looks like a miniature version of the movable nest hive. The upper part of the actual hive could have been made of wood, and the hole on the top could be an effort to prevent overheating. As for the bigger opening, it might have been a passage to enter and exit the hive. In movable nest hives, the honeycombs are attached to the movable top, as already mentioned. Therefore the process of extracting the honey and cleaning the hive would have been greatly facilitated. Another explanation could be that the upper part of the hive was an effort of the potter to imitate the honeycombs, although Marinatos (*idem*) thought that it imitated bees, irrespective of the material it was actually made of. Again, if it was made of wicker or reeds or leaves and wood, the miniature vessel is very similar to the top bar wicker hives from Greece (Crane 1983, fig. 223; N. Crete and Attica).

Extracting the honey and keeping it clean would have been a problem with the pottery object suggested to be a hive by Doumas (1983; above fig. 10. 1). It is a large cylindrical object (height *c.* 82 cm, diameter *c.* 52 cm). At its base, there is a small door, and inside the pot there is a horizontal sieve-like terracotta screen. The top is ridged for receiving a lid, and the body is decorated with crocus flowers. As Begg noted (Hitchcock, unpubl. 1995), the object is more likely to have been used to filter, strain, or even to smoke materials. The sieve in the middle is problematic if the object is to be interpreted as a hive.

Terracotta objects discussed by Hitchcock (*idem*) found at Phaistos (*ca.* 1750-1400 BC), in "Magazine 33" were interpreted as stools or hives. They are made of coarse clay and are 40 cm high, 30-40 cm wide and 3 cm thick. They are open at the bottom, and have two opposite semi-lunate cut-outs on the side. However, almost any pottery cylinder could be used as a top bar, and the arguments used are not convincing.

The miniature hive from Akrotiri might be the predecessor of the top bar hive. The fact that neither of them is mentioned by the classical writers does not mean that they were not known in Antiquity. They may have been out of use during the Classical period, used in a limited area, or the technique was forgotten for a long period. The Akrotiri miniature vessel, together with the Linear B evidence, provides evidence that bees were exploited for honey and wax during the mid second millennium BC in the Aegean.

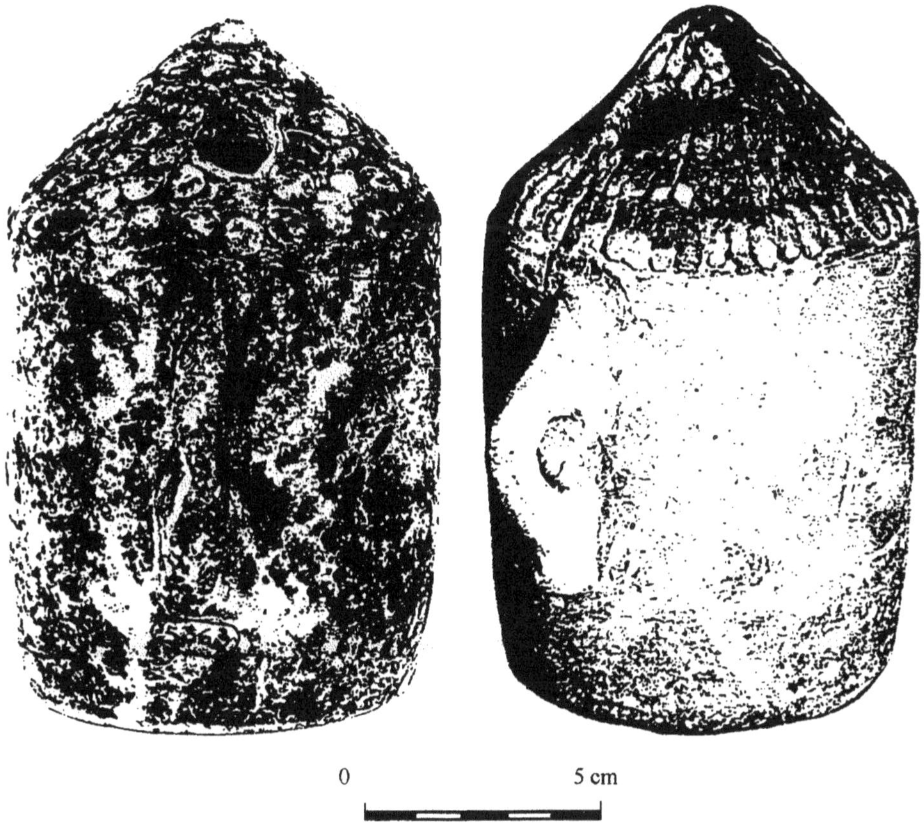

Fig. 10. 4. Model beehives from Xeste 3, Akrotiri.
(after Marinatos 1976).

11. Insect fossils from Egyptian Sites

Introduction

In order to compliment the research undertaken at Akrotiri, and to understand the geographical distribution of insect pests and their inter-relations in the eastern Mediterranean basin, material from Egyptian sites was also been studied. The choice of Egypt (fig. 11. 1) was obvious. Suitable material is not only relatively easily accessible, distributed amongst many museums, but preservation is often optimal. The connections with the Aegean and the rest of the eastern Mediterranean are also obvious. Botanical samples were studied for insect pests from material preserved in the Liverpool Museum (dated to around 1800 BC), and from the settlement associated with the Roman stone quarry at Mons Claudianus (1st Century AD). Coprolites from the Workmen's Village at el Amarna (1400 BC) were also examined.

Egyptian samples from the Museum of Liverpool collections

Examination of several food offering preserved in samples from the Liverpool Museum resulted in evidence of pest infestation. The material was desiccated and the preservation excellent. In most cases, the insects were intact. Unfortunately, some of the samples studied were unprovenanced, and so could not be dated.

Methodology

The material was studied in the Museum, under a stereomicroscope. Identification of the insect species was carried out in Sheffield, with the use of the departmental comparative collection.

Discussion of finds (Table 11.1)

The most exciting material was from XIIth Dynasty Kahun (1900-1800 BC). *Rhyzopertha dominica* (F.), the earliest Egyptian record, was found in barley. *R. dominica* feeds on grain already infested by primary pests, or on mouldy and badly broken grain (Hickin 1968). In *Acacia* seeds from the same site was found one individual of a *Bruchidius* sp., a pest of pulses. Nuts from the site, half eaten by rodents, included an elytron of *Stegobium paniceum* (L.), the biscuit beetle. This may have been a primary pest of the nuts, but could also have been was introduced in mouse faeces. *S. paniceum* was a common find in other samples, including unprovenanced bread cakes from tombs, and also figs and infested lentils.

Taxon	fig seeds Acc. No. 1978-291 224	infested lentils Acc. No. M:C:11122	Kahun (nuts) Acc. No. 56.21. 299	Kahun XII (barley) Acc. No. 56.21. 302	Kahun XII (Acacia)	Unprov bread & cakes
Coleoptera						
Bostrychidae						
Rhyzopertha dominica (F.)				4		
Anobiidae						
Stegobium paniceum (L.)	1	5	1			39
Bruchidae						
Bruchidius sp.					1	
Diptera						1
Puparium indet.						1

Table 11: 1. Insect remains from samples in the Liverpool Museum

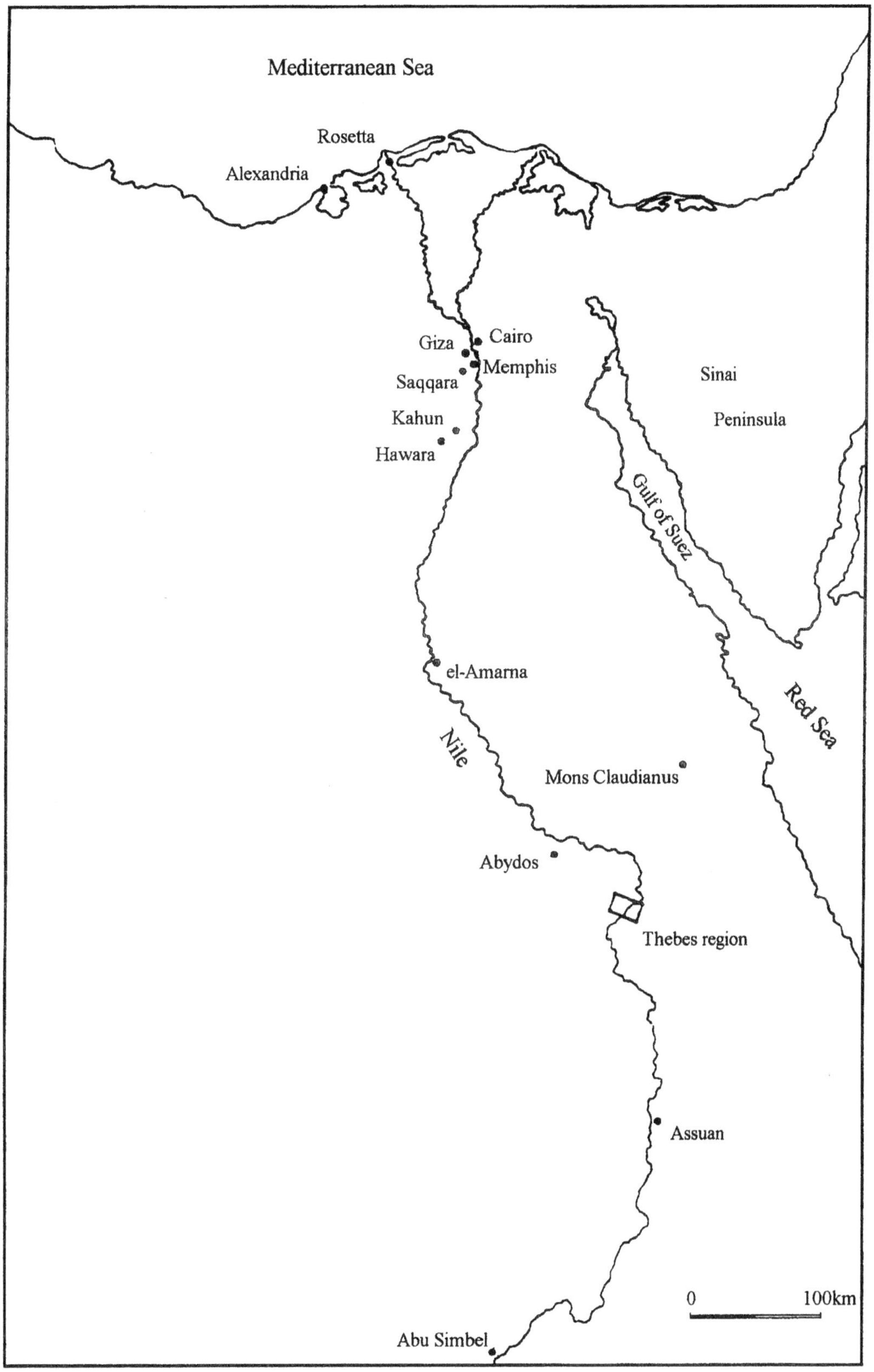

Fig. 11. 1. Egypt : Sites mentioned in the text.

Several other examples of Egyptian or Near Eastern food commodities in the Museum showed evidence of infestation. In most cases the level of infestation was minimal (for example, 3 infested seeds in 150 barley grains from Ptolemaic Hawara). However heavier infestation was also present: of 17 Lotus seeds from XII Dynasty Kahun, 5 were infested, of 209 caryopses of barley and 4 grains of wheat from Kahun, 8 grains of barley were infested (infestation 4%), 25 were infested in 261 grains of barley from XII Dynasty Kahun (infestation 9%), and there were 2 infested seeds in 46 seeds of barley from XII Dynasty Kahun (infestation 4%). In unprovenanced lentils (Museum Accession No.; M: C: 111222) 80 in 609 were infested (13%) and 3% of fig seeds from Ptolemaic Hawara showed evidence of attack.

Carbonised seeds of wheat from Jericho were also examined for infestation. Only 5 seeds in 200 were infested. Unfortunately the wheat was not assigned to any particular period or context. Jericho was excavated by a German team between 1907 and 1909, by Garstang 1930-36 and by Kenyon 1952-58. The site has evidence for occupation from the Mesolithic to the Iron Age (Bray & Trump 1970).

Preliminary Environmental Assessment - Amarna Coprolites

Studies on coprolites from archaeological sites can provide information about diet, hygiene, and the surrounding environment of a site, including the species of animal related to the excrement (e.g. Baker & Brothwell 1980; Hall *et al.* 1983; Bouchet & Paicheler 1995). The Amarna excavation offers a unique opportunity for the study of environmental material, as the preservation of the archaeological remains is excellent as a result of the extreme aridity of the climate in the Egyptian desert.

The site

The Workmen's Village (fig. 11. 2) is isolated from the main city of el-Amarna, by a distance of about 1.2 km. The village was built on the floor of one branch of a shallow Y-shaped valley in a low plateau which runs out from the foot of the cliffs that surround Amarna. The most favoured explanation about the role of the site is that it housed workers and artists employed in the cutting and decorating of the rock-cut tombs (Kemp 1984). Possible involvement in the building of the royal tomb of Akhenaton is complicated by the existence of a second and still unexcavated village lying a kilometre to the east. An alternative explanation has to do with the network of ancient tracks that lead to the desert behind the main city. Many of these can be interpreted as routes for the patrols of the police which were stationed in the city (Davies 1906, 12-18). The village is situated in a place very suitable for a foot patrol system. A further possibility is related to the stratigraphy of the site, and that both explanations are correct: the village was built for tomb workers, abandoned for a while, and then reoccupied in the reign of Tut'ankhamun by a contingent of guards (Kemp 1984). The dates of the Workmen's Village has to be set within the general dating framework for Amarna as a whole. Amarna is related to the reign of Akhenaten; however, the Workmen's Village has to do with a new phase of activities undertaken after Tut'ankhamun had come to the throne. Since the pottery from the Workmen's Village is overall homogenous, and nothing has been found in the village mentioning the name of a king later than Tut'ankhamun, it is likely to have been occupied for only twenty to twenty five years, starting with regnal year 4 of Akhenaten going on for thirteen years, through the short reign of Smenkhare and ending with the nine year reign of Tut'ankhamun (Kemp 1984).

A part of the Workmen's village has been interpreted as animal pens (Shaw 1984), and coprolites from the area have been subject to examination in order to find out the species of animal which were kept. The parasitological analysis (Donald 1984) has revealed eggs of *Ascaris* spp. (either *Ascaris suum* (pig) or *Ascaris lumbricoides* (man)), a worm which infests humans and pigs, as well as *Taenia* spp., the tapeworm genus that lives in man and a range of domesticated and wild animals (Noble & Noble 1976). Taking into account the nature of the coprolites, the ones that produced *Taenia* spp. were amorphous lumps, while the ones that produced *Ascaris* were well formed coprolites, the excavator came to the conclusion that the pens were used to stall pigs (Shaw 1984). A botanical report on the coprolites from the 1980-81 excavations shows that the animals in question were being fed on unthreshed emmer wheat spicelets, barley, rye grass and sedges (Renfrew 1985).

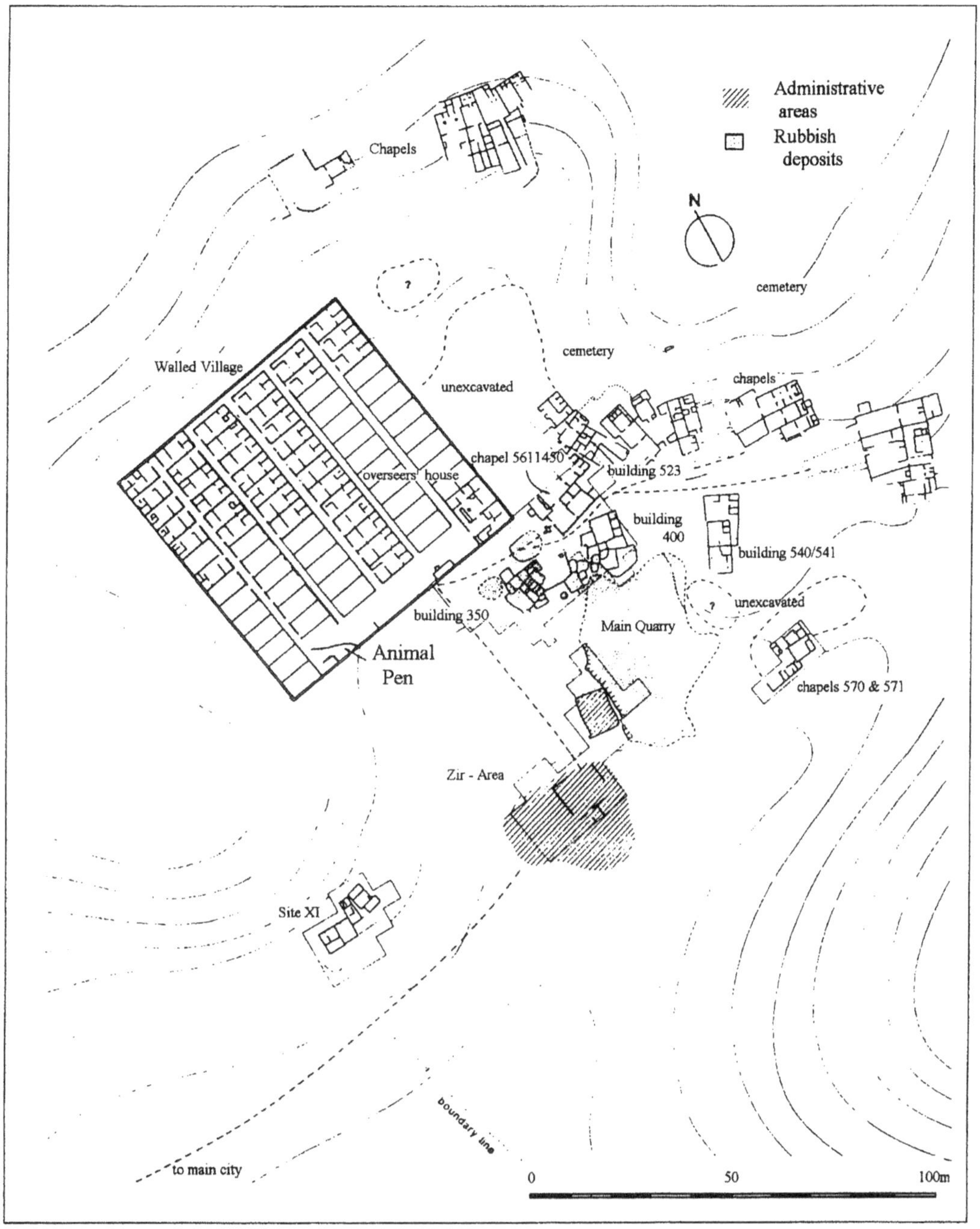

Fig. 11. 2. The Workmen's village, Amarna, showing excavations of 1921-22 and 1979-83 (after Kemp 1985).

The results

Six of the coprolite samples from the Workmen's village were studied under a stereomicroscope. All environmental remains were recorded (Table 11. 2), and plant remains, hair, mammal bones and insect remains were recovered from the processed coprolites. The botanical remains were of varied preservation, and included desiccated, slightly burnt or completely charred plant fragments, consisting mainly of emmer, *Triticum dicoccum*, wheat spikelets and in some cases seeds, barley, *Hordeum vulgare*, and rye grass, *Lolium* sp.

Coprolite sample No.	Plant remains	Hair	Bones	Insect Remains
TA85WV CH541 1914, 1	large amount of plant remains (emmer wheat spikelets + rye grass)	hair (pig)	Small mammal bones	indet. fragments of Dipterous puparia. *Palorus ratzeburgi* (right elytron)
TA85WV Ch541 1914	rye grass and emmer wheat spikelets	hair (pig)	Small mammal bone	indet. fragment.
TA80WV L13, 3	slightly burnt plant remains together with desiccated ones (rye grass).	-	Small mammal bone	*Palorus ratzeburgi* (thorax). Fragment of a Tenebrionidae elytron. fragments and a whole Dipterous puparium
TA80WV P26 Nfac8	charred emmer spikelets and seeds of emmer and barley and + uncharred glumes and chaff			*Sitophilus granarius* (thorax) Cryptophagidae indet. (left elytron) Dipterous puparia fragments
10859 TA83WV R17 406	desiccated plant remains (rye grass)			Coleoptera indet. fragment
TA85.WV?P25 Eface 10	desiccated plant material (rye grass+?flax seed)			

Table 11 : 2. Material recovered from coprolite remains from the Workmen's Village, El Amarna, Egypt

The picture given by the plant remains is very similar to that described by Renfrew (1985). Probably the animals associated with the faeces were fed on wheat waste or, in some cases, bread waste. The latter is indicated by the presence of charred fragments, perhaps from the base of loaves. A few insects were also recovered from the coprolites (Table 11. 2 & 3). *Sitophilus granarius* L., the grain weevil, one of the most serious pests in granaries of the present day, was found in one of the samples studied. *Palorus ratzeburgi* (Wiss.), the small eyed flour beetle, was recorded in two of the samples. *P. ratzeburgi* is a pest in cereal products and other commodities, particularly in mouldy grain residues previously attacked by grain weevils (Brendell 1975). The finds in the Amarna coprolites are the earliest records of the pest. Other species recorded were of the family Cryptophagidae, usually associated with damp and mouldy deposits, often stored products. Tenebrionidae include stored product pests, but many species are associated with arid habitats, and the unidentified fragment from the coprolite probably relates to the natural background fauna of the site. The beetles are closely associated with the plant remains, and provide further evidence for the deduction that part of the diet of the animals consisted of infested cereals. High infestation levels can make the grain inedible, and in some cases, depending on the increase of the microflora, poisonous. Feeding the domestic animals on wheat waste products, as well as infested grains is a common practice still in the Mediterranean (Halstead & Jones 1989, Halstead 1990, Belas pers. comm.).

Taxon	MNI
Coleoptera	
Cryptophagidae	
indet.	1
Tenebrionidae	
indet.	1
Palorus ratzeburgi (Wiss.)	2
Curculionidae	
Sitophilus granarius (L.)	1
Diptera	
Calliphoridae puparia	3
puparium indet.	1

Table 11. 3. Insects from the Amarna coprolites

A few small mammal bones were also recovered from the samples. Largely too fragmentary for identification or undiagnostic, their size range would suggest mice, a not improbable further pest of stored products.

The fly puparial fragments found in the samples are very fragmented, and their identification cannot be certain. However, they most probably are Calliphoridae, the maggots of which feed on carrion, indicating that meat or similar residues were also fed to the animals, a food combination very suggestive of pigs.

Redding (1991) has recently discussed the role of pigs in Ancient Egypt. Pictorial representations are rare, and they are never depicted as offerings to gods (Kemp, pers. comm.). Perhaps they were not considered appropriate food for the Egyptian upper class either. However the maintenance of the pig stalls in the Workmen's Village implies that they were used as food at least for the lower classes.

Mons Claudianus

Introduction

Mons Claudianus is a Roman quarry settlement in the Eastern Desert of Egypt (Fig. 11.3). The environment of the site was not favorable to agriculture, and the largest amount of the diet of the inhabitants had to rely on imported food-stuffs. The plant remains, together with the insect remains, give us further insight in the economy of the site (van der Veen 1996; 1998).

The Site

Mons Claudianus is situated 500km south of Cairo and 120km east of the Nile. The settlement was occupied during the late first and second centuries AD, and was Rome's source for a granodiorite used for columns and decorative stonework during the 1st and 2nd century AD (Peacock 1988; Peacock & Maxfield 1997). The site consists of a fortified settlement, stables, a granary, a well, cemetery, bath house, and associated midden deposits. As a result of the aridity of the region, preservation is excellent, and as well as structures, organic materials and ostraca (Bingen *et al.* 1992) have been recovered from the site.

Methodology

The plant and insect remains of the site were preserved by desiccation. However, charred material was also found in the ovens of the annex and the midden deposits. (van der Veen 1996). The methodology for the recovery of the botanical remains is described in van der Veen (1996). Fractions of several samples were sorted for insects, and the fossils recovered identified under a microscope using a comparative collection. Together with the botanical remains, insects, mostly pests of stored products, were recovered.

The food plants

The botanical analysis (van der Veen 1996; 1998) has given evidence of a varied diet with cereals, pulses, fruits, nuts, herbs and seasonings. Several species of cereals including barley, *Hordeum* sp. and hard wheat, *Triticum durum*, were recovered from the site. Also a variety of pulses such as lentils, *Lens culinaris*, peas, *Pisum sativum*, fava, *Vicia faba* - var. *minor*, termis beans, *Lupinus alba* and chick peas, *Cicer arietinum* were part of the diet. Different kind of nuts were also consumed: almonds, *Amygdalus communis*, pinenut kernels, *Pinus pinea*, hazelnuts, *Corylus avellana*, and walnuts, *Juglans regia*, together with a variety of fruits: dates, *Phoenix dactylifera*, dom fruits, *Hyphaene thebaica*, olives, *Olea europea*, grapes, *Vitis vinifera*, Egyptian plums, *Cordia myxa*, water melons, *Citrullus vulgaris*, bitter apples, *Citrullus colocynthis*, melon or cucumber *Cucumis* sp. pomegranate, *Punica granatum*, Christ's thorn, *Ziziphus spina-christi* and citron, *Citrus medica*. A large variety of herbs, seasonings and vegetables were recovered in the botanical samples. Onion, *Allium cepa*, garlic, *Allium sativum*, fennel, *Foeniculum vulgare*, celery, *Apium graevolens*, coriander, *Coriandrum sativum*, cumin, *Cyminum cyminum*, black cumin, *Nigella sativa* and globe artichoke, *Cyrana* cf. *scolymus* were well known food commodities at the quarry site. In addition, flax, *Linum usitatissimum*, and safflower, *Carthamus tinctorius* were present.

The animal bone analysis has shown that the diet of the settlement included meat of donkey, camel, Dorcas gazelle, presumably hunted locally, eggs, and fish, mostly marine and imported from the Red Sea (Hamilton-Dyer 1990).

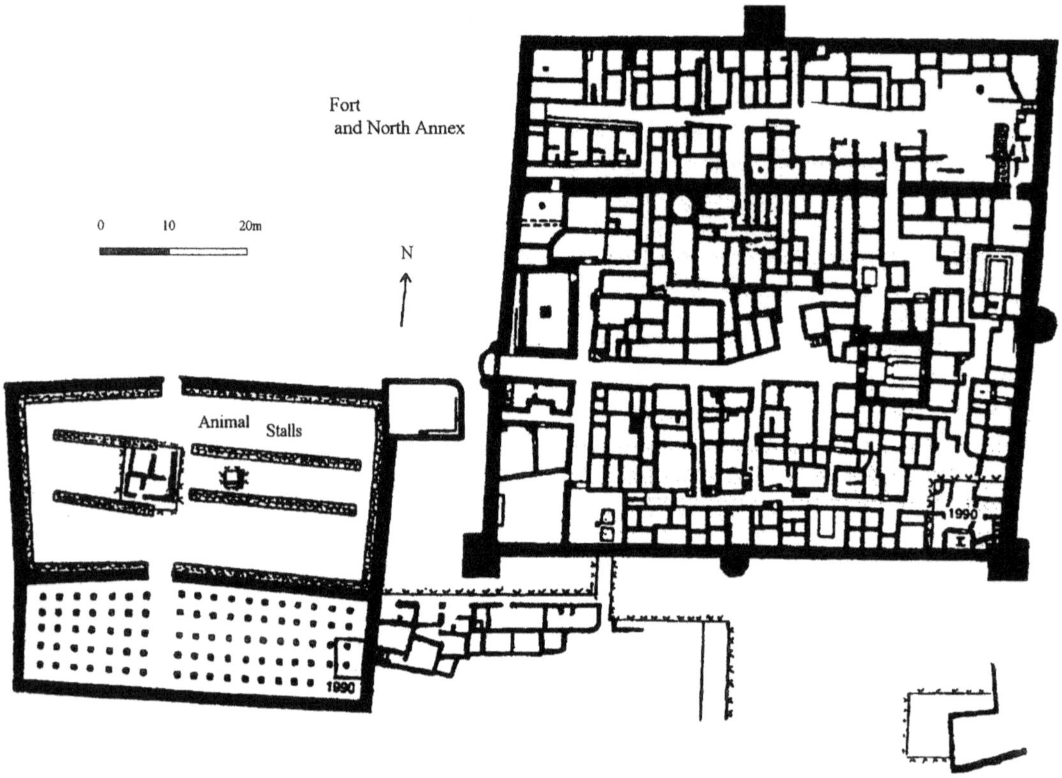

Fig. 11. 3. Plan of the fort and outbuildings, Mons Claudianus, Egypt (after Bingen *et al.* 1990)

The archaeoentomology

The insect faunas from Mons Claudianus are largely typical storeroom faunas (Table 11.4). *Sitophilus granarius* (L.), the grain weevil, a flightless primary pest of stored cereals, lives and develops in grains of corn, rye, maize, oats, buckwheat, millet, and chickpeas (Hoffman 1954). Normally it is transported by Man by shipment of food commodities and in general through trade, it has become cosmopolitan. *Tribolium castaneum* Hbst. is a secondary pest of stored products. It is found together with cereal products in mills, bakeries and rarely in flour mills (Brendell 1975). *Rhyzopertha dominica* (F.) is also a secondary pest in all kinds of starchy food products. Bruchids are mainly pests on pulses, infesting them in the field. The larvae live inside the seeds, that are totally destroyed after their emergence. *Cryptolestes turcicus* Grouv. is not a direct pest, but a mould and detritus feeder. It is found in cereals and their products (Halstead 1993), and the Mons Claudianus examples represent the earliest record.

Gibbium sp. is a synanthropic insect, usually found in dry food stores. The commonest species are *G. psylloides* (Cz.) and *G. aequinoctiale* Boield., until recently confused (Constantine 1994). *Gibbium* is found in houses, hotels, mills, latrines and granaries, where it feeds on grain and vegetable products in general. A specimen of the colydiid *Aglenus brunneus* (Gyll.) was also recovered from the site. *A. brunneus* is currently a rare pest in the wet grain residues, and was probably introduced to stored products when caves where used for food and grain storage (Conn & DeMoss 1984). Records include cellars, stables, under manure, under mouldy straw, among sacks of old bones, in cowsheds, and in mill refuse (Hinton 1945). *Alphitobius diaperinus* (Panz.), the lesser mealworm beetle, was also recovered in the Mons Claudianus samples. It is an omnivorous feeder, associated with a wide range of commodities, e.g. grains, flour, leather, bones, ground nuts, etc. (Brendell 1975). The species has recently been recorded from el-Amarna (Panagiotakopulu, unpubl.).

Overall the insect fauna is dominated by *Rhyzopertha dominica*. Its high numbers, together with *Tribolium castaneum*, imply imports of processed food commodities, such as flour and other milled products.

Taxon	S1	S2	S3	S4	S5	S6	S7	S8
Orthoptera								
Acridoidea indet.								ff
Siphonaptera								
indet.								ff
Coleoptera								
Histeridae								
indet.			1					
Dermestidae								
Anthrenus sp.					1			
Cryptophagidae								
Cryptophagus sp.								1
Cucujidae								
Cryptolestes turcicus (Grouv.)				1				
Colydiidae								
Aglenus brunneus (Gyll.)				2				
Bostrychidae								
Rhyzopertha dominica (F.)	1	6	26	33		2	5	10
Ptinidae								
Gibbium sp.			3	3	4			
Tenebrionidae	1							
Tribolium castaneum (Hbst.)		1	5	5				
Alphitobius diaperinus (Panz.)					1		1	3
Tenebrio sp.				5	1			
indet.								
Bruchidae								
Bruchus sp.				2				
indet.		1	3					
Curculionidae								
cf. *Anthonomus* sp.			1					
Sitophilus granarius (L.)	1		5	9		3	4	12
Diptera								
indet. puparia						2	1	1
Hymenoptera								
Formicidae indet.						5		1

Sample codes :

S1 = MC91 Fort SE corner Room 4 14 I'/W. S2 = MC91 Fort SE corner Room 4 38 R' IW.
S3 = MC92 Annexe S1 1/2 Flot E9. S4 = MC92 Annexe S1 1/4 Flot E2.
S5 = MC90 Fort SE corner A 1/2 R2SW 26. S6 = MC90 Fort S'IE 13B.
S7 = MC90 Fort 6 1/4 S' SW. S8 = MC90 Sebahli 1/4 C7, 5.

Table 11 : 4. Insect remains from Roman Mons Claudianus

The recovery of locust fragments, Acridoidea, together with the insect pests in the material from Mons Claudianus, might be accidental. On the material available, it is not possible to identify the fragments to a closer taxonomic level. Locusts are widespread in the warmer parts of the Palaearctic region, and two species in particular *Schistocera gregaria* L., the desert locust, and *Locusta migratoria* L. (Bei-Bienko 1963, 1964) make periodic migratory excursions into the Mediterranean, and occasionally beyond (Waloff 1946). Locusts are known to have been eaten in the Classical period, the sources being discussed by Beavis (1988; 76-79). Greek sources refer to the insects used for food (Aesopica 387 and Vit. Aes. 99; Galen 17B.484 K; Atheneus 63c). They were, however, mainly eaten either by the poor (Schol. Aristophanes Ach. 116; Scholiast ap. Th.L.L.VII.1605.65), or by soldiers (Aristophanes Ach. 116; Porphyry de Abst. I.25). Also locusts, *locustae* were used to feed young birds. Pliny (HN VII.29) refers to tribes in India that ate locusts. In other places around the Mediterranean, locusts were a well known food. In Cyrene (Porphyry de Abst. I.25) they were a popular dish, while Pliny testifies that an African tribe literally followed locust swarms (HN VII 104). Herodotus IV.172 mentions a Libyan tribe, the Nassamones, that used to put sun-dried and ground locusts in their milk. Also, according to Pliny (HN, VI.195), an Ethiopian tribe was known as Akridophagoi, because the largest part of their diet consisted of *akrides* (= locusts, grasshoppers). Diodorus describes how the locusts were caught; fires were lit at the bottom of a ravine where the locusts passed by during spring; the insects fell to the ground, affected by the smoke. Afterwards, the locusts were collected, and made into savoury cakes (Jerome MPL 24.364). Bodenheimer (1951, 139, 160) claims that these locusts were *S. gregaria* and *L. migratoria*, which are used for food today (Beavis 1988).

Among Greeks, singing insects were often captured and kept as pets (Beavis 1988; 76). There is evidence for cicadas, but it is mainly *akrides* that were kept this way in the "ακριδοθήκη", a cage for locusts made of plants or reeds (Theocritus I.52-3, Longus I.10). Mention of such a cage exists also in the Anthology, and finally an epitaph by Anyte (AP VII.190=VII.364) testifies that *akris* and *tettix* (cicada) were kept as pets (Beavis 1988;103).

Fragments of Siphonaptera, fleas, were also recovered from Mons Claudianus, although they could not be identified to the species. There are different species of fleas, all parasites, living on the blood of mammals and birds, some restricted to a single host, others infesting a range of species (Rothschild & Clay 1952). Female fleas do not attach their eggs to the hair or feathers of the animal; either they leave the animal and lay their eggs in organic refuse, or they lay their eggs while on the host and they are scattered in the nest or sleeping place. Measures against flea infestation consist to a variety of mixtures of herbs boiled up in water and sprinkled on the infested area (Pliny HN XX 172, XXII 27, 49, XXIV 53; Galen XIV 537 K; Dioscorides DMM.III.121, IV.15; Ps. Apuleius Herb. 103.4; Geoponica XIII.15.2-3, 5-6). Herbal fumigants (Pliny HN XX 155), lime (Geoponica XIII.15.7) and different concoctions and magical remedies were also applied (e.g. Pliny HN XXX 85) (Beavis 1988).

Conclusions

The botanical remains found in Mons Claudianus indicate commodities imported, from the Nile valley. The existence of a variety of insect pests, related to food commodities provides a classic example of introduction of insect pests to different parts of the world through shipments and trade (*cf.* Buckland 1981). The Romans while expanding their Empire, or establishing quarries in remote areas for economic or military purposes, not only introduce their culture, but also different insect pests that travelled together with food commodities destined for the soldiers or the workmen. The insect fossil assemblage from Mons Claudianus is a unique record of dispersal of insects associated with stored products dependent upon trade.

12. Trade in perishables

"Lo, all His Majesty's harbours are supplied with all good things of His Majesty's booty from Djahi, Keftiu-ships, Byblos ships, and skt-ships of s-sood being loaded with poles, floorboards together with large timber for his Majesty's great timber yards."

Annal of Thutmose III. Temple of Amon at Karnak, Egypt.
XVIII Dynasty. 34th year of Tuthmose III's reign.

(Strange 1980;74-75 no.22).

Introduction

In the Early Bronze Age, during the Cretan Protopalatial period, there is evidence for links between the Aegean, Egypt and the Levant (Knapp 1989). The main body of information consists of finds of inorganic goods, ceramics, metals and stone, which are always better preserved in the archaeological record and tend to focus all attention in discussions of trade in the eastern Mediterranean (Knapp 1989). However, it is evident that other commodities were also transported around the Mediterranean basin, often in significant quantities. Although they do not preserve well in the archaeological record, or, in some cases, at all, such materials were clearly important in the economies of the region. The evidence is scattered, but a wide range of organic goods, such as food commodities (grain, vegetables, fruits, as well as wine, beer, etc.), plants, trees, reeds and rushes, and all their by-products - textiles, garments, drugs, dyes and timber, incense, resins, basketry, animal or fish products (leather/hide, fats, oils), spices, honey, and wax are all recorded as having been traded around the Mediterranean basin in Antiquity, but have left little archaeological trace (Knapp 1989).

Evidence for trade in perishable materials

Limited archaeological data and evidence from the literary record witness widespread trade in perishable goods from the Middle Bronze Age onwards in the Mediterranean.

Chemical analysis has been employed increasingly to identify residues. The use of vegetable oils was ascertained by analysis of the contents of a Late Minoan I pot, the so-called "Abbot jug", (Merrillees & Winter 1972, 106-108;130), and wine was suggested for the contents of the Late Minoan I "Marseilles ewer" (Merrillees & Evans 1980, 17). Bitumen, salt and animal fat were recorded from the Late Minoan I "Alternating style cup" (*idem*, 32-36), and olive oil occurred in two amphorae and a Gallo-Roman lamp (Condamin *et al.* 1976). Dye substances have also been recovered, as, for example, components extracted from the madder plant *Rubia tinctorum* L., proven by analysis of Iron Age Cypriot White slip pottery (Foster & Moran 1989).

Occasionally from the archaeological evidence, it is evident that commodities were being exported from the Aegean to Cyprus, Egypt and the Levant. The excavation of the Late Bronze Age shipwreck at Ulu-Burun, off the south coast of Turkey, provides substantial evidence of the shipment of raw materials. From the shipwreck were recovered tons of copper, tin, ingots of glass, ebony, a ton of terebinth resin, hippopotamus teeth, part of an elephant tusk, tortoise shells, orpiment, ostrich eggshells, *Murex* shells, spices and foodstuffs (Bass 1991). The mix of substances suggests a coasting vessel taking up trade items from a number of locations. An "exotic" legume, claimed to be a definite import from the Aegean, was found at the Bronze Age site of Tel Nami, near Haifa, Israel (Kislev *et al.* 1993). Charred seeds of *Lathyrus clymenum* L. were found in situ in four storage jars, and on the floors of two storerooms dated to the Middle Bronze IIA period, *ca.* 1950-1750 BC. *Lathyrus clymenum* today is a rare and local food plant in the western Mediterranean countries. It is an uncommon pulse crop due to toxic substances in the seeds which, if eaten in large quantities for a long time can cause a serious disease, called lathyrism, or neuro-lathyrism. The plant is still cultivated on some Aegean islands, and was found at the site of

Akrotiri (Sarpaki 1987). The climate of Israel is said not suitable for its cultivation, and it was probably imported form the Aegean (*idem*, 145-154).

The Keftiu, Cretan, bean is recorded in the Papyrus Ebers (1600 BC):

> "Another (remedy) for giving
> defecation; *gngnt6* (An unknown herb),
> it will be like the bean from Keftiu."
>
> (Vercoutter 1956, 39; Cline 1994).

Its name implies that the species of pulses were exported from Crete or the Aegean, before any local cultivation in Egypt (Warren 1995). In view of the Tel-Nami find, the Keftiu bean can be identified as *Lathyrus clymenum*, Kislev and others (1993), however, believe that the term is a reference to some other type of pulse.

Indirect evidence for trade of perishables also comes from the Linear B tablets (1400-1300 BC). Production of olive oil, wine, honey, fruits, cereals, as well as textiles and perfumes are recorded. The output of the textile and perfume industries would suggest production beyond subsistence, and thus the existence of surplus goods for export (Knapp 1989, 42; Shelmerdine 1985, 147-148; Killen 1985, 266-267; 272-273). Certain types of fabrics and oils are defined as *ke-se-ne-wi-jo/-ja* (*xenwios/a*: guest welcoming), which could be associated with "external trade" or "trading exchanges". Also *ku-pi-ri-jo* could mean: "of Cypriote quality" or "suitable for export to Cyprus" (Knapp 1989, 42).

An Akkadian text from Ugarit on the Syrian coast, east of Antioch, reveals that Crete was drawn into the Levantine orbit of maritime trade in organic goods ((PRU 3, 107 (RS 16.238) in Knapp & Cherry 1994). The merchant Sinaranu was granted duty free status on grain, a fermented beverage, and oil that he had imported from Crete (Kabduri (Caphtor) Liverani (1979, 1330) quoted in Bass (1991)). An inventory of ceramics and luxury goods from Mari, on the right bank of the Euphrates near the Syrian Iraqi border, (ARM 24, 98;10,14) also lists Cretan (Kaptara) products including fabrics (Dossin 1939, 111-112; Maxwell-Hyslop 1970). Cuneiform documents state that one hundred ships from Ugarit transported 2000 measures of grain to Cilicia during the 13th century BC (Otten 1967, 59; Heltzer 1978, 210; Breasted 1906, 3; 580). In another letter (PRU 6.14), two Ugaritic merchants pondered the merit of voyages to Egypt, and of shipping to Hatti (Anatolia) bronze tools, garments and mules (Knapp & Cherry 1994, 145; Courtois 1990).

Painted motifs on garments pictured in Egyptian tombs from Thebes, are identified as Aegean, for example, the ones of Antef, Amenemhet, Hapuseneb and Menkheperreseneb (1900-1300 BC, in Barber 1991). Syrians and 'Keftiu' from the processions depicted in the tombs wear clothes that are of very similar fashion, and textiles are depicted in the tomb of Menkheperreseneb as one of the gifts brought by the embassy (fig. 12. 1) generally interpreted as Cretan. Barber (1983) believes that Aegean women weavers were working at Kahun, during the XIIth Dynasty (1900-1800 BC). Minoan style wall paintings have been found at Tell el Dab'a in the Nile Delta, in contexts indicating a late Hyksos and an early XVIIIth Dynasty date (Middle - Late Bronze Age (Bietak 1995; Morgan 1995)). A plastered floor and walls decorated in Minoan style are known from Middle Bronze Age II levels at Tel Kabri in Israel (Neimeier 1991; Collon 1994; Negbi 1994). The stone-built, barrel-vaulted, so-called Mycenaean tombs at Ugarit (Schaeffer 1933, 117-18; 1939, 77-97; Evans 1935, 776), together with Late Helladic IIA-IIB pottery found in Cyprus, Ugarit, on the Levantine coast, and in Egypt (Schaeffer 1939, 72-76; Muhly 1970, 44, 48; Åstrom 1973; Buchholz 1974; Courtois & Courtois 1978, 292-363; Bell 1982), show the inter-relationships between people of the Mediterranean basin, and the division between the Aegean and Anatolia / Syria is perhaps as much an artifact of recent political prejudices as an actuality.

The picture presented from the Ulu Burun shipwreck, as well as from the literary record, the artistic representations and the archaeological record, is that of close interrelations between the Aegean, Cyprus, Egypt and the Levantine coast. Clearly ideas of thalassocracies in the basin are completely inappropriate. The trade networks and relations during the Late Bronze Age in the Mediterranean, were complex and influenced by politics (Cline 1994, 107). Bass (1967, 75-78) argues that Minoans and Syrians played a pre-eminent role in Mediterranean trade.

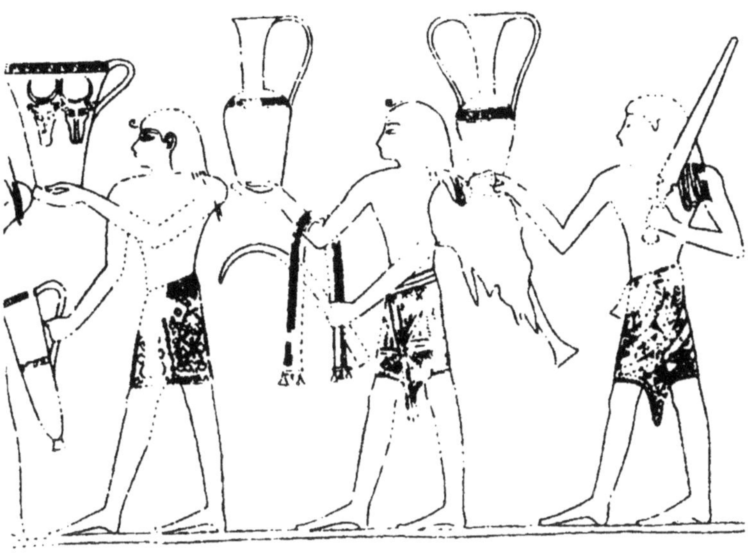

Fig. 12. 1 : Detail from the Tomb of Menkeheperraseneb (Th.86), mid-15th Century B.C. One of the Aegean ambassadors holds a bolt of cloth over his arm (after Davies & Davies 1933, pl. 5).

Yannai (1983, 110-114) believes in Levantine middlemen who served as regional factors in supply and demand. This is based on the fact that there is a wide range of Levantine materials to be found in the Aegean, whilst there is a lack of Mycenaean exports in the Levant. Based upon literary and historical evidence, Yannai also argues for indirect trade in finished products and raw materials redistributed by Levantine middlemen based in Cyprus.

During the Bronze Age, the Levantine city states had well established political traditions, ruled by hereditary kings, and the Egyptian political hierarchy made no attempt to change their organisation. Tributes were a standard form of exchange, while the military or naval presence of the Egyptians was mostly symbolic (Liverani 1971, 268). Archives, like the Amarna letters, a contrast with the Royal inscriptions which were addressed to the Egyptians, aiming to portray the superiority of the Pharaohs, demonstrate the reciprocal nature of East Mediterranean / Aegean trade (Knapp & Cherry 1994). Warren has recently argued (1995, 14) that the "Minoan" civilization derived benefit from Egypt in many different forms, but only a very small part of the traded items are likely to be recovered in the archaeological record. Many, if not most items were perishable goods, and it cannot be ascertained whether they came from Egypt (Fig.13.2), Syro-Palestine or even Mesopotamia. Ports, for example, Ugarit in Syria, Kommos on Crete, and Enkomi in Cyprus, were probably important trading centres during the Bronze Age (Knapp & Cherry 1994), involved, as the Ulu-Burun wreck suggests, in the extensive reshipment of commodities.

Insects as indicators of trade

The role of insects as indicators of trade and movement of goods, has been discussed at length by Buckland (1981; 1991), and a significant amount of new data has accumulated since his work. Particularly in the Old World, the distribution of storage pests implies trading mechanisms and exchange of commodities. The ways in which insect pests moved from their original natural habitats into the storerooms, and from there were transported all around the world has already been considered. As most pests appear to have originated in the warmer parts of the world, the Mediterranean offers ideal climatic conditions, and the storerooms an optimal living environment to provide reservoirs for shipment throughout the basin after initial introduction.

In some cases, when the primary item of trade has perished, the chitinous cuticle of the insects related to the commodity may survive to testify to its existence. Insect fossils may be the only indicators of products and activities which, if not for them, would go unnoticed. In the case of economic insects, for example silkmoths, bees, or insects related to dyestuffs, the recovery of the fossils witnesses facets of the economy that would not be traced otherwise. An example has been provided by the find of the calcified cocoon from Akrotiri. Vital information on wild silk cultivation in the Aegean during the Late Bronze Age has come to light and given a new dimension to our perception of the economy of that period.

When interpreting artistic evidence the element mostly taken into account is the style of the artist. Terms as Minoan, Aegean or Theran style are widely used (Warren 1995; Bietak 1995; Morgan 1995), and most scenes tend to be interpreted as cult scenes or religious ceremonies. The interpretation of wall paintings, seals, rings and pottery is filtered through this traditional point of view. However, other factors should be considered. Specifically, the artistic representations from the Aegean, are characterised by a strong naturalistic element, and on most occasions, the artists depict quite accurately scenes of everyday life. Therefore a pragmatic approach to artistic representations provides valuable information on the reconstruction of the relevant period. In the case of the Akrotiri wall paintings, the geographical distribution of the depicted animals, reveals links with Egypt and the Levant. By linking the animals of the wall paintings to the archaeology and natural history, sometimes the mystical aspect of the scene is eliminated or at least reduced, and interpretation reflects more the economic importance of the connections. For example, when the butterflies in the prows and masts of the ships represented in the flotilla fresco (West House, Room 5, South Wall) are not interpreted as references to deities but as silkmoths and the flower next to them as cotton, the wall painting acquires a completely different character. Whilst a ceremonial element is still evident, this bears no relation to the reconstruction of the scene as a procession of battleships (Doumas 1992).

An important consideration related to the trade in luxury items (e.g. silk and dyestuff) is the way they acquire such a high profile. Also, another interesting point is the mechanisms devised by the elite in order to control their distribution. Local elites may have generated their position of power from controlling certain types of goods; exploitation and dominance may be results of trade (Godelier 1977). Sherratt and Sherratt (1991) and Manning (1994) have argued that during the Bronze Age in the Mediterranean the most significant incentive for trade was the desire of local elites to acquire specific high-status goods. The import of certain types of goods required the intensification of the production of other, raw or finished materials, in order to exchange the commodities, as well as the spread of a certain lifestyle that generated demand and consumption (Knapp & Cherry 1994). Examined in this light, trade in wild silk, cotton, dyestuffs, wax and honey and exotic foodstuffs during the Late Bronze Age in the Aegean could have been a powerful tool for gaining control of the circulation of goods. In the Aegean, Akrotiri probably played an important role as a redistribution centre in low-bulk, high value commodities, whilst its own requirements may have included the import of grain, a pathway by which insect pests of cereals are likely to have arrived.

In the case of Egypt, the picture is much more complex, with trade from Africa, south of the Sahara, India and Arabia via the Red Sea ports and Europe and the Near East all potentially identifiable on their accidentally introduced insect faunas.

13. Conclusion

Summary of results

Preservation of insect remains varies due to a range of different factors. Both wet and arid environments may produce almost perfectly preserved assemblages. Disaggregation over a 300µm sieve followed by paraffin (kerosene) flotation has become the standard technique for samples recovered from damp anaerobic conditions, and wet sieving, over a range of different sieve sizes has been employed for the recovery of charred botanical remains, usually from dry sediments. The latter has been found to be appropriate for insect recovery from desiccated assemblages, although carbonized insect fossils may disintegrate in water. Dry sieving and hand sorting of samples, a time consuming and little rewarding method, is the recommended technique for processing of charred material which includes insect remains, at least until a new method is found. The differential preservation of charred insect chitin observed during the study of the various fossil assemblages implies varied resistance to high temperatures and/or different conditions of carbonisation, and more research is required on this aspect.

The recovery of a variety of synanthropic insects in the Late Bronze Age settlement of Akrotiri, Thera, Greece, implies trade contacts involving perishable materials between the Aegean and Egypt and the Near East. Some of the species recovered were spread around Europe only subsequently, when the Roman army dispersed them along with food provisions (Buckland 1981). Others, such as the Dermestids, appear not to have been established outside the Mediterranean before the seventeenth century AD. Further research on insect fossils from Egyptian sites, - XIIth Dynasty Kahun (1900-1800 BC), the *ca.* 1400 BC. Workmen's Village at Amarna, 1st - 2nd century AD Mons Claudianus - has highlighted the size of the problem of insect infestation in ancient Egypt, and also provided additional evidence for the dispersal of pests through trade.

Different storage techniques since the beginnings of agriculture if not before, have been used to control and counteract the problem of pest infestation. Most interestingly, the continuous use of plants and other substances with insecticidal properties since prehistory around the Mediterranean has been established (Panagiotakopulu *et al.* 1995).

The evidence for wild silk cultivation in relation to finds related to textile making, such as loom weights and probable use of dyes on site, suggests that the manufacture of luxurious textiles could have been a primary occupation in the settlement of Akrotiri. The pictorial evidence from the wall paintings at Akrotiri as well as from Minoan rings, seals and wall paintings of the same period attest the use of diaphanous garments, probably made of wild silk, in the Late Bronze Age Aegean. The wild silk moth, *Saturnia pyri*, until recently interpreted as a butterfly, is depicted in the Akrotiri wall paintings, and is a frequent artistic theme in the Cyclades, Crete, and mainland Greece during the Late Bronze Age. Further evidence from the Akrotiri wall paintings implies knowledge and probable use of the cotton plant for cloth manufacture. Also, beekeeping was perhaps practised by the inhabitants of Akrotiri, as shown by the finds of two miniature beehive models.

The overall picture of Akrotiri, based on the insect finds, is of a flourishing settlement, extensively involved in the trade in food commodities and perhaps finished luxury goods, including fine cloth. This was strongly linked with the Levant and Egypt.

Further research

An interest by North Europeans in the archaeology of the Eastern Mediterranean dates back at least to the eighteenth century, when wealthy travellers collected and took back with them as memorabilia archaeological finds that ranged from art objects to mummified bodies. Since then, much has happened. But for a long time, archaeological research in the area of the basin had either a colonial or even a nationalistic face that not only limited its scope but still creates problems today.

During the latter half of the twentieth century, at the time that Mediterranean archaeology was struggling to obtain a professional character, archaeological science was becoming more widely practised in the northern parts of the world. As the relative paucity of finds on archaeological sites in the British Isles, Scandinavia or Iceland for example called for a more detailed approach, scientific research has proven to be an excellent tool for reconstructing the past. In the Mediterranean on the other hand, the publication of numerous archaeological sites could best be described as a 'Marie Celeste situation' - everything is placed on the table, but there is nobody home. The description of the more impressive finds, as for example, architectural remains and pottery, have always been a priority for the archaeologist, while environmental research has only lately started entering the agenda.

Apart from the fact that finds were until recently looked upon as entities and not in relation to the context from which they came, there has also been a tendency to examine archaeological sites and archaeological material from a very localised point of view. However, environmental factors attribute similar characteristics to areas around the Mediterranean and up to a point determine their growth and development. As much as during the initial stages of study, archaeological material has to be looked at in the framework of the locality, it also has to be set in the wider context, the whole area of the basin.

To add even more complications, where the traditional approach and the lack of field and laboratory research stops, the grand models start. But theories with no substantial research behind them have no detail, and such lack of detail can support numerous hypotheses. Insect infestation is a detail that is hardly ever mentioned in economic models. Whilst not apparent to the archaeologists, it can have affected drastically a community's everyday life. Models are the definition of food for thought in archaeology: they should result from research, and originate even more research. Armchair philosophers with no understanding of the region, do the same for archaeological study as the art collectors of the past century.

Ideally the way of undertaking archaeological research in the Mediterranean should be by researching the environment, past and present, understanding the ethnography and applying archaeological science in order to interpret archaeological sites and finds. Among the traditonal approaches and the fragmentary, undigested use of science, a few but substantial efforts have made a difference. Kemp and others (1994) epitomise the use of archaeological research for reconstructing everyday life in Egypt, where previously the abundant documentary record has lead to a neglect of environmental aspects of research and where the numerous lists of plant remains rarely contain any substantive interpretation.

From an entomological point of view, the Aegean islands and littoral, Egypt, the Levant and the rest of the area, were occupied by endemic and anthropochorous insect faunas, closely related. They were also occupied by trading populations with similar lifestyles and related economies. These aspects have a close relationship with Mediterranean archaeology, and many more sites require investigation before an overview of human impact upon insect faunas and insect impact upon human societies can be achieved.

References

Aelian (1958). *On the characteristics of animals.* (trans. A. F. Scholfield). Loeb Text, Heinemann, London.

Aitkin, A. D. (1975) *Insect Travellers. 1. Coleoptera.* Ministry of Agriculture, Fish & Food, Agricultural Development & Advisory Service, Pest Infestation Control Laboratory, HMSO, London.

Alexander, P., Kitchener J. A. & Briscoe, H. V. A. (1944). Inert dust insecticides. *Annals of Applied Biology* **31**, 143-159.

Alfieri, A. (1931). Les insectes de la tombe de Toutankhamon. *Bulletin de la Société Royale Entomologique d 'Égypte*, **3/4**, 188-189.

Alfieri, A. (1976). *The Coleoptera of Egypt.* Atlas Press, Cairo.

Allchin, B. & Allchin, R. (1982). *The rise of civilisation in India and Pakistan.* Cambridge University Press.

Alluaud, C. (1908). Notes sur les coléoptères trouvés dans les momies d' Égypte. *Bulletin de la Société Royale Entomologique d' Égypte*, **1**, 29-36.

Amouretti, M. C. (1979). Les cereales dans l' Antiquité: éspeces, mouture, conservation, liaisons et interferences. In M. Gast & F. Sigaut (eds.) *Les Techniques de Conservation des Grains à Long Terme,* I, 57-69. Centre Nationale de la Recherche Scientifique Paris.

Andres, A. (1931). Catalogue of the Egyptian Tenebrionidae. *Bulletin de la Société Royale Entomologique d' Égypte* **15**, 74-125.

Anon. (1945). *Greece, III. Regional Geography.* Naval Intelligence Division, London.

Anon. (1978). The British Museum's beeswax treasures. *Bee World* **59**(1), 39-40.

Anon. (1989). *The Periplous Maris Erythraei* (trans. by L. Casson). Princeton University Press.

Aristophanes (1924). *Lysistrata* (trans. by B. B. Rogers). Loeb Text, Heinemann, London.

Aristotle (1910). *Historia Animalium* (trans. J. Smith & W. Ross). Oxford University Press.

Arkell, A. J. (1971). Comment on "Origins of food production in southwestern Asia" by G. A. Wright. *Current Anthropology* **12**, 471-472.

Aristotle (1936). *Minor Works* VI (trans. W. S. Hett). Loeb Text, Heinemann, London.

Ashe, H. B. (1941-55) (ed.) *Columella, Lucius Iunius Moderatus. De re rustica.* Loeb Text, Heinemann, London.

Åstrom, P. (1973). Comments on the corpus of Mycenaean pottery in Cyprus. In V. Karageorghis (ed.) *Acts of the International Archaeological Symposium: the Mycenaeans in the Eastern Mediterranean,* 122-127. Zavallis Press, Nicosia.

Atti, R. & Kamel, A. H. (1965) The fauna of stored products in U.A.R. *Bulletin Société entomologique d'Égypte* **49**, 221-232.

Audouin, J. V. (1835) Observation sur la découverte de *Gibbium scotias* dans un ancien tombeau à Thébes. *Bulletin Société Entomologie de France* **5**, 5-6.

Baeta Naeves, C. M. (1971). Survey of continental storage conditions in Portugal and storage conditions in the Portugese islands (Madeira and Azores). *Tribolium Information Bulletin* **14**, 66-75.

Baillie, M. G. L. (1990). Irish tree rings and an event in 1628 BC. In D. A. Hardy (ed.) *Thera and the Aegean World III,* **3**, 160-166. The Thera Foundation, London.

Baillie, M. G. L. (1998). Bronze Age myths expose archaeological shortcomings? a reply to Buckland *et al.* 1997. *Antiquity* **72**, 425-427.

Baillie, M. G. L. & Munro, M. A. R. (1988) Irish tree rings, Santorini and volcanic dust veils. *Nature* **322**, 344-346.

Bain, A. L. (1997) *Archaeoentomological analysis at the Ferryland site, Newfoundland.* Memorial University of Newfoundland Archaeology Unit, St. Johns.

Bain, A. L. (1998) A seventeenth century beetle fauna from colonial Boston. *Historical Archaeology*, **32**, 38-48.

Bainbridge Fletcher, T. & Ghosh, C. C. (1921). Storage receptacles in use among the people. In T. Bainbridge Fletcher & C. C. Ghosh, Stored Grain Pests. *Bulletin of the Pusa Agricultural Research Institute* **111**, 713-758 (reprinted in Sigaut 1978, 126-153).

Baker, J. & Brothwell, D. (1980). *Animal Diseases in Archaeology,* 172-188. London, Academic Press.

Baker, A. S. (1990). Two new species of *Lardoglyphus* Oudemans (Acari: Lardoglyphidae) found in the gut contents of human mummies. *Journal of stored Product Research* **26**, 139-147.

Bandy, A. C. (1983). (trans.) *Ioannes Lydus, De magistratibus rei publicae romanae.* The American Philosophical Society, Philadelphia.

Barber, E. J. W. (1991). *Prehistoric Textiles.* Princeton University Press.

Barrett, J. (1996). *Few know an Earl in fisherman's clothes.* Unpubl. PhD dissertation, University of Glasgow.

Bass, G. F. (1967). Cape Celidonya: A Bronze Age Shipwreck. *Transactions of the American Philosophical Society,* **57**.

Bass, G. F. (1991). Evidence of trade from Bronze Age shipwrecks. In N. H. Gale (ed.) *Bronze Age Trade in the Mediterranean. Studies in Mediterranean Archaeology,* **90**, 69-82. Åstrom, Goteborg.

Beavis, I. C. (1988). *Insects and other invertebrates in Classical Antiquity.* University of Exeter, Exeter.

Becker, E. C. (1978). Dermestoidea. In H. V. Danks (ed.) *Canada and its Insect Fauna.* Memoir of the Entomological Society of Canada, **108**, 375-376.

Begg, D. J. (1975). *Minoan storerooms in the LBA.* Unpubl. PhD dissertation, University of Toronto, Canada.

Behrensmeyer, A. K., & Kidwell, S. M. (1985). Taphonomy's contributions to palaeoecology. *Palaeobiology* **11**(1), 105-119.

Bei-Bienko, M. (1963). *Locusts and grasshoppers of the USSR and adjacent countries* **I**. Israel Program for Scientific Translations. Jerusalem.

Bei-Bienko, M. (1964). *Locusts and grasshoppers of the USSR and adjacent countries,* **II**. Israel Program for Scientific Translations. Jerusalem.

Bell, M. R. (1982). Preliminary report on the Mycenean pottery from Deir el Medina. *Annales du Service des Antiquités de l' Égypte* **68**, 142-163.

Bennett, E. L. (1950). Fractional quantities in Minoan book-keeping. *American Journal Archaeology* **54**, 204-222.

Betancourt, P. P. & Weinstein, G. A. (1976). Problems of Middle Bronze Age chronology: New evidence from Egypt. *American Journal of Archaeology* **88**, 471-485.

Betts, A., van der Borg, K., de Jong, A., McClintock, C. & van Strydonck (1994) Early cotton in North Arabia. *Journal of Archaeological Science,* **21**, 489-499.

Betts, J. (1984). The sealstones and sealings. In M. Popham *et al.* (eds.), *The Minoan unexplored mansion,* Text: 187-196. Thames & Hudson, London. British School at Athens. Supplementary Volume **17**.

Bietak, M. (1995). Connections between Egypt and the Minoan World: New Results from Tell el-Dab'a /Avaris. In W. V. Davies & L. Schofield (eds.) *Egypt, the Aegean and the Levant. Interconnections in the Second Millennium BC,* 19-28, British Museum Press, London.

Bingen, J., Búlow-Jacobsen, A., Cockle, W., Cuvigny, H., Rubenstein, L. & Rengen, W. van (1992) *Mons Claudianus. Ostraca Graeca et Latina* **I**. (*O. Claud. 1 á 190*). Documents de Fouilles, Institut Français d'Archéologie Orientale du Caire **29**, Cairo.

Blaisdell, F. E. (1927). A blind beetle excavated from an Egyptian city's ruins dating between 117 and 235 A.D. *Proceedings of the Entomological Society of Washington* **29**, 121-125.

Bodenheimer, F. S. (1928-29). *Materialen zur Geschichte der Entomologie bis Linne.* **I-II**. Junk, Berlin.

Bodenheimer, F. S. (1951). *Insects as Human Food.* The Hague.

Boessneck, J. (1988) *Die Tierwelt des alten Ägypten, untersucht anhand kulturgeschichtlicher und zoologischer Quellen.* C.H.Beck, Munich.

Borchardt, L. (1916). Das ältagyptische Wohnhaus im 14. Jahrhundert v. Chr. *Zeitschrift fur Bauwesen.* (Berlin) **66**, 510-558.

Bottema, S. (1980). Palynological investigations on Crete. *Review of Palaeobotany and Palynology* **31**, 193-217.

Bouchet, F. & Paicheler, J. C. (1995). Presumption of a case of bilharzia on an archaeological site of the XVth century at Montbellard (Doubs, France). *Comptes Rendues de l' Academie des Sciences Serie III-Sciences de la Vie-Life Sciences,* **318**, (7), 811-814.

Bradley, P. L. (1958). An assemblage of arthropod remains from a Roman occupation site at St. Albans. *Nature* (London), **181**, 435-436.

Bray, W. & Trump, D. (1970). *A Dictionary of Archaeology*. Penguin, Harmondsworth.

Breasted, J. H. (1906). *Ancient Records of Egypt*, **I-V**. University of Chicago Press, Chicago.

Brendell, M. J. D. (1975). *Coleoptera: Tenebrionidae. Handbooks for the identification of British Insects, V,10*. Royal Entomological Society of London.

Bresc, H. (1979). Fosses a Grains en Sicile (XIIeme-XVeme siecles). In M. Gast & F. Sigaut (eds.) *Les Techniques des Conservation de Grains a Long Terme*, I, 113-122. Centre National de la Recherche Scientifique, Paris.

Brothwell, D. R. (1969). *Food in Antiquity, a survey of the diet of early peoples*. Thames and Hudson, London.

Brouwer, P. (1967). *General Palaeontology*. Oliver & Boyd, Edinburgh.

Brunner-Traut, E. (1975). Affe, *Lexikon der Agyptologie*. Wiesbaden I, 83-85.

Buchelos, C. T. & Katopodis, C. A. (1995). A survey of beetles in store-rooms containing barley and maize for animal feed, on the Island of Lefkas, Greece: abundance and population fluctuation of the most significant species. *Journal of stored Product Research,* **31**, 3, 253-258.

Buchholz, H. G. (1974). Agäische Funde und Kultureinflusse in den Randgebeiten des Mittelmeers. Forschungsbericht uber ausgrabungen und Neufunde. *Archaologisches Anzeiger* 325-462.

Büchner, S. & Wolf, G. (1997) Der Kornkäfer - *Sitophilus granarius* (Linné) - aus einer bandkeramischen Grube bei Göttingen. *Archäologisches Korrespondenzblatt*, **27**, 211-220.

Buckland, P. C. (1976a). *The environmental evidence from the Church Street Roman sewer system*. Archaeology of York, **14/1**. Council for British Archaeology for York Archaeological Trust.

Buckland, P. C. (1976b). *Niptus hololeucus* Fald. (Col., Ptinidae) from Roman deposits in York. *Entomologist's monthly Magazine*, **111**, 233-234.

Buckland, P. C. (1978). Cereal production, storage and population: a caveat. In S. Limbrey & J. G. Evans (eds.) *The effect of Man on the Landscape : the Lowland Zone*, 43-45. Council for British Archaeology, Research Report **21**, London.

Buckland, P. C. (1981). The early dispersal of insect pests of stored products as indicated by archaeological records. *Journal of stored Product Research* **17**, 1-12.

Buckland, P. C. (1982). The Malton burnt grain: a cautionary tale. *Yorkshire Archaeological Journal* **54**, 53-61.

Buckland, P. C. (1990) Granaries Stores and Insects. The Archaeology of Insect Synanthropy. In D. Fournier & F. Sigaut (eds.) *La préparation alimentaire des céréales,* 69-81. Rapports présentés à la Table ronde, Ravello au Centre Universitaire pour les Biens culturels, avril 1988. PACT, Rixensart, Belgium.

Buckland, P. C. (1996) Insects. In J. May, *Dragonby, Report on Excavations at an Iron Age and Romano-British settlement in North Lincolnshire*, 165-170. Oxbow Monograph **61,** Oxford.

Buckland, P. C. (1997) Review of Kenward & Hall (1995). *Journal of Archaeological Science*, **24**, 962-963.

Buckland, P. C. & Coope, G. R. (1991). *A Bibliography and Literature Review of Quaternary Entomology*. J. R. Collis Publications. Department of Prehistory and Archaeology, Sheffield University.

Buckland, P. C., Dugmore, A. J. & Edwards, K. J. (1997). Bronze Age Myths? volcanic activity and human response in the Mediterranean and North Atlantic region. *Antiquity* **71**, 581-593.

Buckland, P. C., Dugmore, A. J. & Edwards, K. J. (1998). Comment. *Antiquity* **72**, 431-432.

Buckland, P. C., Sadler, J. P. & Sveinbjarnardóttir, G. (1992) Palaeoecological investigations at Reykholt, Western Iceland. In, C. J. Morris & D. J. Rackman (eds.) *Norse and later settlement and subsistence in the North Atlantic,* 149-168. Dept. of Archaeology, University of Glasgow.

Buckland, P. I., Yuan Zhuo, D. & Buckland, P. C. (1997) Towards an expert system in palaeoentomology. In A. C. Ashworth, P. C. Buckland & J. P. Sadler (eds.) *Studies in Quaternary entomology - An inordinate fondness for insects. Quaternary Proceedings* **5**, 67-78.

Burleigh, R. & Southgate, B. J. (1975). Insect infestation of stored Egyptian lentils in antiquity. *Journal of Archaeological Science* **2**, 391-392.

Bury, R. G. (1929). (trans.) *Plato Timaeus; Critias; Creitophon; Menexenus; Epistles*. Loeb Text, Heinemann, London.

Byl, S. (1980). *Recherches sur les grands traites biologiques d'Aristotle: sources écrites et préjugés*. Paris.

Caesar, Julius (1955) *Alexandrian, African and Spanish Wars.* (A.G. Way trans.) Loeb Text, Heinemann, London.

Camps, G. (1961). *Aux origines de la Berberie. Monuments et rites funéraires préhistoriques*. A. M. G, Paris.

Capasso, L. & Ditota, G. (1998) Lice buried under the ashes of Herculaneum. *Lancet*, **351**, 992.

Carrington-Smith, J. (1975) *Textile manufacture in prehistoric Greece*. Unpubl. Ph.D., University of Tasmania.

Carter, D. J. & Hargreaves, B. (1986) *Caterpillars of Britain and Europe*. Harper Collins, London.

Caseldine, A. (1987) Charcoal patches from Meare Village East 1982. In J. M. Coles, Meare Village East. The excavations of A. Bulleid and H. St. George Gray 1932-1956. *Somerset Levels Papers* **13**, 223-226.

Cato the Censor (1933). *On Farming* (trans. E. Brehaut). Columbia University Press, New York.

Casson, L. (1989). (trans.) *The Periplous Maris Erythraei*. Princeton University Press.

Cato, Marcus Porcius (1935). *On Agriculture*, (trans. W. D. Hooper). Loeb Text, Heinemann, London.

Chaddick, P. R. & Leek, F. F. (1972). Further specimens of stored products insects found in ancient Egyptian tombs. *Journal of stored Product Research*, **8**, 83-86.

Chadwick, J. (1958). *The Decipherment of Linear B*. Cambridge University Press, Cambridge.

Chitere, P. O. & Omolo, B. O. (1993). Farmers' indigenous knowledge of crop pests and their damage in western Kenya. *International Journal of Pest Management* **39** (2), 126-132.

Chowdury, K. A. & Buth, G. M. (1971). Cotton seeds from the neolithic in Egyptian Nubia and the origin of Old World cotton. *Biological Journal of the Linnaean Society* **3**, 303-12.

Chu, H. F. & Wang L.-Y. (1975). (Insect carcasses unearthed from Chinese antique tombs). *Acta Entomologica Sinica*, **18**, 333-337 (In Chinese)

Cline, E. H. (1994). *Sailing the Wine-Dark Sea. International trade and the Late Bronze Age Aegean*. British Archaeological Reports **S591**. Oxford.

Cockburn, A., Barraco, R. A., Reyman, T. A. & Peck, W. H. (1975). Autopsy of an Egyptian mummy. *Science*, **187**, 1155-1160.

Collon, D. (1941). Bull leaping in Syria. *Agypten und Levante* **4**, 81-88.

Columella (1941-55). *Res Rusticae*, (trans. by H. B. Ash), I-IV. Loeb Text, Heinemann, London.

Columella (1954-55). *Res Rusticae*, (trans. by E. S. Forster & E. N. Heffner), V-XII, (2 vols.). Loeb Text, Heinemann, London.

Condamin, J., Formenti, F., Olmetalis, M., Michel, M., & Blond, P. (1976). The application of gas chromatography to the tracing of oil in ancient amphorae. *Archaeometry* **18**, 195-202.

Conn, D. B. & De Moss, G. L. (1984). Distribution of four troglophilic beetles in a *Myotis sodalis* (Chiroptera) hibernaculum. *Coleopterists Bulletin*, **38**, 251-255.

Constantine, B. (1994). A new ecological niche for *Gibbium aequinoctiale* Boieldieu (Ptinidae) in Britain, and a reconsideration of literature references to *Gibbium* spp. *Coleopterist* **3**, 25-27.

Conway, S. (1992). *Thai textiles*, 61-84. British Museum Press.

Coope, G. R. (1994) The response of insect faunas to glacial - interglacial climatic fluctuations. *Philosophical Transactions of the Royal Society of London* **B344**, 19-26.

Coope, G. R. (1995) The effects of Quaternary Climatic changes on insect populations: lessons from the past. In R. Harrington & N. E. Stork (eds.) *Insects in a changing environment*, 29-48. Academic Press, London.

Coope, G. R. & Osborne, P. J. (1968). Report on the coleopterous fauna of the Roman well at Barnsley Park, Gloucestershire. *Transactions of the Bristol and Gloucestershire Archaeological Society* **86**, 84-87.

Cooper, E. K. (1962) *Silkworms and science.* Lutterworth, London.

Costantini, L., Tosi, M. & Vigna-Taglionti, A. (1977). Typology and socio-economical implications of entomological finds from some ancient Near Eastern sites. *Paleorient*, **3**, 247-258.

Courtois, J. C. (1990). Yabninu et le palais sud d' Ugarit, niveau historique. Deuxieme partie. *Ugaritica 7. Mission de Ras Shamra* **18**, 191-370. Geuthner, Paris.

Courtois, J. C. & Courtois, L. (1978). Corpus ceramique de Ras Shamra-Ugarit, niveau historique. Deuxieme partie. *Ugaritica 7.* Mission de Ras Shamra **18**, 191-370. Geuthner, Paris.

Crainge, M. & Ahmed, S. (1988). *Handbook of plants with pest-control properties.* New York.

Crane, E. (1983). *The archaeology of beekeeping.* Duckworth, London.

Currid, J. & Navon, A. (1989). Iron Age pits and the Lahav (Tell Halif) grain storage project. *BASOR* **273**, 77-78.

Curry, A. (1979). The insects associated with the Manchester Mummies. In A. R. David (ed.) *Manchester Museum Mummy Project,* 113-117. Manchester University Press, Manchester.

Davidson, D. A. (1978). Aegean soils during the second millennium B.C. with reference to Thera. In C. Doumas (ed.) *Thera and the Aegean World I,* 725-739.

Davies, N. de G. (1906). *The rock tombs of El Amarna,* **IV**. London.

Davies, N. de G. & Davies, N. (1933). *The Tomb of Menkheperrasonb Amenmose, and Another*, London.

Davies, M. & Kathirithamby, J. (1986). *Greek Insects.* Duckworth, London.

Davis, E. N. (1992). A storm in Egypt during the Reign of Ahmose. In D. A. Hardy, J. Keller, V. P. Galanopoulos, N. C. Flemming & T. H. Druitt (eds.) *Thera and the Aegean World III*, **3**, 232-235. The Thera Foundation, London.

Delobel, A. & Malonga, P. (1987). Insecticidal properties against *Caryedon serratus* (Ol.) (Coleoptera: Bruchidae). *Journal of stored Product Research* **23**, 173-176.

de Moulins, D. (1990) Environmental Analysis. In C. Mahony & D. de Moulins, *The Archaeology of Roman London* **1**. *The Upper Walbrook in the Roman Period,* 85-152. Council for British Archaeology Research Report **69**. London.

Dendy, A. & Elkington, H. D. (1920). Report on the effect of air-tight storage upon grain insects. *Grain Pests Committee Report* **6**, Royal Society of London.

Denford, S. (1978). Mites and their potential use in archaeology. In Brothwell, D. R., Thomas, K. D., & Clutton-Brock (eds.) *Research problems in zooarchaeology.* Institute of Archaeology of London Occasional Papers **3** (3), 77-83.

Diapoulis, S, C. (1980). Prehistoric plants of the islands of the Aegean sea, sea daffodils (*Pancratium Maritimum*). In C. Doumas (ed.), *Thera and the Aegean World II* , 129-140. The Thera Foundation, London.

Dillon, D. S. & Dillon, L. S. (1972). *A manual of common beetles of Eastern North America* (2 vols.). Dover Publications, New York.

Dio Chrysostom (1940). (trans. J. W. Cohoon & H. L. Crosby). Loeb Text, Heinemann, London.

Dobie, P., Haines, C. P., Hodges, R. J., & Prevett, P. F. (1984). *Insects and Arachnids of Tropical Stored Products: Their Biology and Identification.* Slough, Tropical Development and Research Institute.

Donald, C. R. (1984) Examination of the archaeological samples supplied by Barry J. Kemp from Tell el-Amarna site. In B. J. Kemp, *Amarna Reports* **I**, 56-58. Egypt Exploration Society, London.

Don-Pedro, K. N. (1987). *Insecticidal activity of plant oils against stored product pests.* Unpubl. PhD thesis, University of London.

Donisthorpe, H. ST. J. K. (1939). *The Coleoptera of Windsor Forest.* Publ. privately, London.

Dossin, G. (1939). Les archives economiques du palais de Mari. *Syria* **20**, 97-113.

Doumas, C. (1974). The Minoan eruption of the Santorini volcano. *Antiquity* **48**, 110-115.

Doumas, C. (ed.) (1978). *Thera and the Aegean World I. Papers and proceedings of the Second International Scientific Congress, Santorini, Greece, August 1978.* Thera and the Aegean World, London.

Doumas, C. (ed.) (1980). *Thera and the Aegean World II. Papers and Proceedings of the Second International Scientific Congress, Santorini, Greece, August 1978.* Thera and the Aegean World, London.

Doumas, C. (1983). *Thera. Pompeii of the ancient Aegean. Excavations at Akrotiri 1967-1979*. Thames & Hudson, London.

Doumas, C. (1992). *The wall-paintings of Thera*. The Thera Foundation, Athens.

Douskos, I. (1980). The crocuses of Santorini. In C. Doumas (ed.), *Thera and the Aegean World II*, 141-146. The Thera Foundation, London.

Duhoux, Y. (1988). Les contacts entre Mycéniens et Barbares d'après le vocabulaire du Linéaire B. *Minos* **23**, 75-83.

Ebbell, B. (1937). *The Papyrus Ebers*. Oxford University Press, London.

Einarson, B. & Link, G. K. K. (1990) (trans.) Theophrastus, *De causis plantarum* Loeb Text, Heinemann, London.

El-Nahal, A. K. M., Schmidt, G. H. & Risha E. M. (1989). Vapours of *Acorus calamus* oil - a space treatment for stored product insects. *Journal of stored Product Research* **25** (4), 211-216.

Ericksen, U., Friedrich, W. L., Buchardt, B., Tauber, H. & Thomsen, M. S. (1990). The Stronghyle Caldera: geological, palaeontological and stable isotope evidence from radiocarbon dated stromatolites from Santorini. In D. Hardy, J. Keller, V. P. Galanopoulos, N. C. Flemming & T. H. Druitt (eds.) *Thera and the Aegean World III*, **2**, 139-150. The Thera Foundation, London.

Evans, A. J. (1928). *The Palace of Minos* II. Macmillan, London.

Evans, A. J. (1930). *The Palace of Minos* III. Macmillan, London.

Evans, A. J. (1935). *The Palace of Minos* IV. Macmillan, London.

Fagerstrom, I. A. (1964). Fossil communities in palaeoecology: their recognition and significance. *Bulletin of the Geological Society of America* **75**, 1197-216.

Fairclough, H. R. (1966). (trans.) *Horace, Satires, Epistles and Ars Poetica*. Harvard University Press, Heinemann.

Field, N. H., Matthews, C. L., & Smith, I. F. (1964). New Neolithic sites in Dorset and Bedfordshire, with a note on the distribution of Neolithic storage pits in Britain, *Proceedings of the Prehistoric Society* **30**, 352-381.

Flannery, K. V. (1971). Origins and ecological effects of early domestication in Iran and the Near East In S. Struever (ed.) *Prehistoric Agriculture*, 50-100. New York.

Forbes, R. J. (1956). *Studies in Ancient Technology* IV. E. J. Brill, Leiden.

Forbes, W. T. M. (1930). The silkworm of Aristotle. *Classical Philology*, **25**, 22-26.

Forster, E. S. & Heffner, E. N. (1955). (trans.) *Columella, Res Rusticae*, VI-XII. Loeb Text, Heinemann, London.

Foster, G. V. & Moran, P. J. (1989). Plants, paints, and pottery: identification of madder pigment on Cypriot ceramic ware. In Y. Maniatis (ed.) *Archaeometry: Proceedings of the 25th International Symposium*, 183-191. Elsevier, Amsterdam.

Foster, J. (1995) Metalworking in the British Iron Age : The evidence from Weelsby Avenue, Grimsby. In, B. Raftery, V. Megaw & V. Rigby (eds.) *Sites and sights of the Iron Age. Essays on fieldwork and Museum research dedicated to Ian Mathieson Stead*, 49-60. Oxbow Monograph **56**, Oxford

Fouqué, F. A. (1879). *Santorin et ses eruptions*. G. Masson, Paris.

Fowler, W. (1886-91). *The Coleoptera of the British Isles* (5 vols.). Nathaniel Loyd, London.

Frank, T. (1940). *The Edict of Diocletian. An economic survey of ancient Rome* V. London.

Fraser, H. M. (1931). *Beekeeping in antiquity*. University Press, London.

Frazer (1913) *Pausanias' Description of Greece*.

Freiser, R. (1981). U. Fam. Otiorhynchinae. In H. Freude, K. W. Harde, & G. A. Lohse (eds.), *Die Käfer Mitteleuropas*, Goecke & Evers, Krefeld. **10**, 184-241.

Friedrich, W. L. (1978). Fossil plants from Weichselian interstadials, Santorini. In C. Doumas (ed.) *Thera and the Aegean World I*, 741-753. The Thera Foundation, London.

Friedrich, W. L. (1980). Larvae of the olive whitefly. *Aleurolobus* (Aleurodes) *olivinus* Silvestri. In W. L. Friedrich, Fossil Plants from Weichselian Interstadials. Santorini (Greece) II. In C. Doumas (ed.) *Thera and the Aegean World II*, 121-122. London.

Friedrich, W. L., Pichler, H. & Kussmaul, L. (1977). Quaternary pyroclastics from Santorini, Greece and their significance for the Mediterranean palaeoclimate. *Bulletin of the*

Geological Society of Denmark **26**, 27-39.

Freude, H., Harde, K. W., & Lohse, G. A. (eds.) (1963-89). *Die Käfer Mitteleuropas*. Goecke and Evers, Krefeld.

Galanopoulos, A. G. & Bacon, E. (1969). *Atlantis*. London & New York.

Gamble, C. (1978). The Bronze Age animal ecology from Akrotiri: a preliminary analysis. In Doumas C. (ed.), *Thera and the Aegean world*, **I**, 745-753. The Thera Foundation, London.

Georghiou, G. (1977). *The insects and mites of Cyprus*. Kiphissia, Athens, Greece.

Gilbert, A. S. & Singer, B. H. (1982). Reassessing zooarchaeological quantification. *World Archaeology* **14**, 21-40.

Gillow, J. & Barnard, N. (1991). *Traditional Indian textiles*. Thames and Hudson, London.

Gimbutas, M. (1974). *The gods and goddesses of Old Europe*. Thames & Hudson, London.

Giordan, J. C. (1988). Sur la description de deux éspeces du *Troglorhynchus* Schmidt (Coleoptera: Curculionidae). *Entomologiste* **44**, 107-109.

Girling, M. A. (1984). Investigations of a second insect assemblage from the Sweet Track. *Somerset Level Papers* **10**, 79-91.

Girling, M. A. (1989). The insect fauna of the Roman well at the Cattlemarket. In A. Down (ed.), *Chichester excavations*, **6**, 234-241. Philimore, Chichester.

Glanville, S. P. K. (1942). *The legacy of Egypt*. Oxford Clarendon Press.

Godelier, M. (1977). Salt money and the circulation of commodities among the baruya of New Guinea. *Perspectives in Marxist Anthropology*, 127-151. Cambridge University Press, Cambridge.

Godley, A. D. (1923-38) (trans.) *Herodotus, Histories* (4 vols.) Loeb Text, Heinemann, London.

Golob, P. & Webley, D. J. (1980). The use of plants and minerals as traditional protectants of stored products. *Tropical Products Institute Report*, **G138/32**. London.

Good, I. (1995). On the question of silk in pre-Han Eurasia. *Antiquity* **69**, 959-968.

Graham, A. J. (1975). Beehives from Ancient Greece. *Bee World* **56**, 2, 64-75.

Graham, A. J. (1978). The Vari House - an addendum. *Annual of British School of Archaeology in Athens* **73**, 99-101.

Greuv, B. (1990). The geographic distribution of Lamprosomatinae, Eumolpinae, Chrysomelinae, Alticinae, Hispinae and Cassidinae in Greece (Coleoptera, Chrysomelidae). *Deutsche Entomologische Zeitschrift* **37**, 289-359.

Grove, K. (1995) A study on insect remains in archaeological deposits. In *The Archaeology of Chichester and District 1995*, 39-40. Southern Archaeology & Chichester District Council.

Gsell, S. (1929). *Histoire ancienne de l' Afrique du Nord*. Hachette, Paris.

Hall, A. R., Jones, A. K. G. & Kenward, H. (1983). Cereal bran and human faecal remains from archaeological deposits - some preliminary observations. In Proudfoot B. (ed.), *Site, Environment and Economy*, 85-104. British Archaeological Reports **S173**. Oxford.

Hall, A. R. & Kenward, H. K. (1990). Environmental evidence from the Colonia. *Archaeology of York*, **14/6**. Council for British Archaeology for York Archaeological Trust.

Hall, A. R., Kenward, H. K. & Williams, D. (1980). Environmental evidence from Roman deposits in Skeldergate. *Archaeology of York*, **14/3**. Council for British Archaeology for York Archaeological Trust, London.

Hall, A. R., Kenward, H. K., Williams, D. & Greig, J. R. A. (1983). Environment and living conditions at two Anglo-Scandinavian sites. *Archaeology of York*, 14/4. Council for British Archaeoogy for York Archaeological Trust, London.

Hall, D. M. & Hyde, M. B. (1954). The modern method of hermetic storage. *Tropical Agriculture* **31**, 149-160.

Halstead, D. G. H. (1963). Coleoptera: Histeridae. *Handbooks for the Identification of British Insects* **IX, 10**. Royal Entomological Society of London.

Halstead, D. G. H. (1993). Keys for the identification of beetles associated with stored products- II. Laemophloieidae, Passandrinae and Silvanidae. *Journal of stored Product Research* **29**, 99-195.

Halstead, P. (1981). From determinism to uncertainty: social storage and the rise of the Minoan palace In A. Sheridan & G. Bailey (eds.) *Economic*

Archaeology, 187-213. British Archaeological Reports **S96**, Oxford.

Halstead, P. (1987). Traditional and Ancient Rural Economy in Mediterranean Europe: Plus ça change? *Journal of Hellenic Studies* **107**, 77-87.

Halstead, P. (1990). Waste not, want not: traditional responses to crop failure in Greece. *Rural History* **1**, 147-164.

Halstead, P. (1992). Agriculture in the Bronze Age. Towards a model of a palatial economy. In B. Wells (ed.) Agriculture in Ancient Greece. *Skrifter utgivna av Svensks Institutet i Athen* **4, XLII. 22**, 105-118.

Halstead, P. & Jones, G. (1989). Agrarian ecology in the Greek islands: time stress, scale and risk. *Journal of Hellenic Studies* **109**, 41-55.

Hamilton-Dyer, S. (1990). The faunal remains. In Peacock, D. P. S. & Maxfield, V. A. (eds.) *Archaeological Reports from Mons Claudianus*. Southampton.

Hansmann, C. & Hansmann, L. (1959). *Viel kostlich Wachsgebild.* F. Bruckmann, Munich.

Harde, K. W. (1984). *A field guide in colour to beetles*. Octopus Books, London.

Hardy, D. A., Keller, J., Galanopoulos, V. P., Flemming, N. C. & Druitt, T. H. (eds.) (1990). Thera and the Aegean World III, **2, 3**. The Thera Foundation, London.

Hardy, P. G., & Aston, P. G. (1990). The Pre-Minoan landscape of Thera: a preliminary statement. In D. A. Hardy, J. Keller, V. P. Galanopoulos, N. C. Flemming & T. H. Druitt (eds.) *Thera and the Aegean World III*, **2**, 348-361. The Thera Foundation, London.

Harlan, J. R. (1992). *Crops and Man* (2nd ed.). Crop Science Society of America.

Harris, J. R. (1971). *The legacy of Egypt.* Clarendon Press, Oxford.

Havry, J. (1905). (trans.) *Procopius, De Bello Gothico*. Leipzig.

Hay, R. (1972). *The Gardening Year.* Reader's Digest, London.

Hedervari, P. (1978). Energetical calculations concerning the Minoan eruption of Santorini. In C. Doumas (ed.) *Thera and the Aegean World* **I**, 179-194. The Thera Foundation, London.

Heiken, G. & McCoy, F. (1990). Precursory activity to the Minoan eruption, Thera, Greece. In, D. A. Hardy, J. Keller, V. P. Galanopoulos, N. C. Flemming & T. H. Druitt *Thera and the Aegean World III,* **1**, 79-88. The Thera Foundation, London.

Heltzer, M. (1978). *Goods, prices and the organization of trade at Ugarit.* Riechert, Wiesbaden.

Herford, G. M. (1935). A key to the members of the family Bruchidae (Col.) of economic importance in Europe. *Transactons of the Society of British Entomology* **2**, 1-32.

Herodotus (1923-38). *Histories* (trans. by A. D. Godley) (4 vols.) Loeb Text, Heinemann, London.

Hesiod (1929). *The homeric hymns and homerica* (trans. H. G. E. White). Loeb Text, Heinemann, London.

Hesychius (1953). *Hesychii Alexandrini Lexicon.* (K. Latte, ed.).

Hickin, N. E. (1963). *The Insect factor in wood decay*. Hutchinson, London.

Hill, D. S. (1975). *Agricultural insect pests of the Tropics and their control.* Cambridge University Press, Cambridge.

Hinds, H. E. (1914). *Agricultural Experimental Station Bulletin*, **176**.

Hinton, H. E. (1945). *A monograph of the beetles associated with stored products*, **I**. British Museum (Natural History), London.

Hinton, H. E. (1948). A synopsis of the genus *Tribolium* MacLeay with some remarks on the evolution of the species-groups (Coleoptera, Tenebrionidae). *Bulletin of entomological Research* **39**, 13-55.

Hirth, F. (1966). *China and the Roman Orient.* New York.

Hitchcock, L. A. (unpubl.) *Of bar stools and beehives. An interpretative dialogue About a Minoan storeroom.* Unpublished paper presented at the 1995 Theoretical Archaeology group meeting.

Hodge, P. J. & Jones, R. A. (1995). *New British beetles. Species not in Joy's practical handbook.* British Entomological and Natural History Society, Reading.

Hoffman, A. (1954). Coléoptères Curculionides. *Faune de France*, **59**. Lechevalier, Paris.

Homer (1980). *The Odyssey* (trans. by W. Shewring). Oxford University Press.

Hood, S. (1976). The Mallia gold pendant: wasps or bees? In Emmison & Stephens (eds.) *Tribute to an antiquary: essays presented to Marc Fitch by some of his friends*. London.

Hooper, W. D. (1935) (trans.) *Marcus Porcius Cato, On Agriculture*. Loeb Text, Heinemann, London.

Hooper, W. D. (1967) (trans.) *Terentius Varro, On Agriculture*. Loeb Text, Heinnemann, London.

Hope, F. W. (1834). Notice of several species of insects found in the heads of Egyptian mummies. *Proceedings of the Royal entomological Society of London*, (1834), 11-13.

Hope, F. W. (1842) Observations on some mummied beetles taken from the inside of a mummied ibis. *Transactions of the Entomological Society of London*, **3**, 191-193.

Hopf, M. & Zacharie, G. (1971). Determination of botanical and zoological remains from Ramat Matred and Arad. *Israel Exploration Journal*, **21**, 60-64.

Horace (1927). *The Odes and Epodes* (trans. by C. E. Bennett). Loeb Text, Heinemann, London.

Horace (1966). *Satires, Epistles and Ars Poetica* (trans. by H. R. Fairclough). Loeb Text, Heinemann, London.

Horion, A. (1956). *Faunistik der Mitteleuropäischen Käfer* **V**. *Heteromera*. Munich.

Horion, A. (1957). *Faunistik der Mitteleuropäischen Käfer*. **VI**: *Lamellicornia (Scarabaeidae-Lucanidae)*. A. Freyel, Überlingen-Bodensee.

Horion, A. (1961). *Faunistik der Mitteleuropäischen Käfer*, **VIII**, *Clavicornia*. Uberlingen-Bodensee.

Housley, R. A., Hedges, R. E. M., Law, I. A. & Bronk, C. R. (1990). Radiocarbon dating by AMS of the destruction of Akrotiri. In D. A. Hardy , J. Keller, V. P. Galanopoulos, N. C. Flemming & T. H. Druitt (eds.), *Thera and the Aegean World III*, **3**: 207-215. The Thera Foundation. London.

Hort, A. (1916) Theophrastus, *Historia Plantarum*. Loeb Text, Heinemann, London.

Howe, R. W. (1955). Studies on Beetles of the Family Ptinidae. 12. The Biology of *Tipnus unicolor* (Pill. & Mitt.). *Entomologist's monthly Magazine* **91**, 253-257.

Howe, R. W. (1965a). A summary of estimates of optimal and minimal conditions for population increase of some stored product insects. *Journal of stored Product Research* **1**, 177-184.

Howe, R. W. (1965b). *Sitophilus granarius* (L.) (Coleoptera Curculionidae) breeding in acorns. *Journal of stored Product Research* **1**, 99-100.

Hubbard, R. N. L. B. (1979). Appendix 2: Ancient agriculture and ecology at Servia. In C. Ridley and K. A. Wardle (ed.) Rescue excavations at Servia 1971-1973. A Preliminary Report. *Annual Report of the British School at Athens* **74**, 226-229.

Hunter, F. A., Tulloch, J. B. M. & Lambourne, M. G. (1973). Insects and mites of maltings in the East Midlands of England. *Journal of stored Product Research* **9**, 119-141.

Hyman, P. S. (1992). *A review of the scarce and threatened Coleoptera of Great Britain*, Part 1 (Revised & updated by M.S.Parsons). UK Joint Nature Conservation Committee, Peterborough.

Ifantidis, M. D. (1983). The movable-nest hive - forerunner of the movable comb hive. *Bee World* **64**, 79-87.

Immerwahr, S. A. (1985). A possible influence of Egyptian art in the decoration of Minoan wall painting. In, P. Darcque and J. C. Poursat (eds.) *L' iconographie minoenne*, 41-50. BCH Suppl. **XI**, Athens.

Ioannes Lydus (1983). *De magistratibus rei publicae romanae*, (trans. A. C. Bandy). The American Philosophical Society, Philadelphia.

Jashemski, W. F. (1979). Pompeii and Mount Vesuvius, A.D. 79. In P. D. Sheets & D. K. Grayson (eds.) *Volcanic Activity and Human Ecology*, 587-622. Academic Press, London

Jeannell, R. (1926). *Faune cavernicole de la France*. Paris.

Jones, A. H. M. (1986). *The Later Roman Empire*. **I-II**, 284-602. Oxford University Press.

Jones, G. (1983). The LM II Plant Remains. In M. R. Popham (ed.), *The Minoan Unexplored Mansion at Knossos*, 303-306. British School at Athens. Thames & Hudson. London.

Jones, G., Wardle, K., Halstead, P., & Wardle, D. (1986). Crop storage at Assiros. *Scientific American* **254**, 96.

Jones, J. E. (1976). Hives and honey of Hymettus. Beekeeping in Ancient Greece. *Archaeology* **29**, 80-91.

Jones, R. E. (1986). *Greek and Cypriot Pottery: a review of scientific studies*. Fitch Laboratory Occasional Papers **1**. Athens.

Joubert, P. C. (1966). Field infestations of stored-product insects in South

Africa. *Journal of stored Product Research* **2**, 159-161.

Karali-Yiannakopoulou, L. (1990). Sea shells, land snails and other marine remains from Akrotiri. In D. A. Hardy (ed.) *Thera and the Aegean World III*, **2**, 410-415. The Thera Foundation, London.

Kaszab, A. (1969). Familie Oedomeridae - Boridae. In H. Freude, K. W. Harde & G. A. Lohse (eds.) *Die Käfer Mitteleuropas* 8, 75-264. Goecke & Evers, Krefeld.

Kemp, B. J. (1984). *Amarna Reports* **I**. Egypt Exploration Society, London.

Kemp, B. J. (1985). *Amarna Reports* **II**. Egypt Exploration Society, London.

Kemp, B. J., Samuel, D. & Luff, R. (1994) Food for an Egyptian city : Tell el-Amarna. In, R. Luff & P. Rowley-Conwy (eds.) *Whither environmental archaeology?*, 133-170. Oxbow Monograph **38**, Oxford.

Kenna, V. E. G. (1960). *Cretan Seals: with a catalogue of the Minoan Gems in the Ashmolean Museum*. Oxford.

Kenward, H. K. (1975). The biological and archaeological implications of the beetle *Aglenus brunneus* (Gyllenhall) in ancient faunas. *Journal of Archaeological Science* **2**, 63-69.

Kenward, H. K. (1976). Further archaeological records of *Aglenus brunneus* (Gyll.) in Britain and Ireland, including confirmation of its presence in the Roman Period. *Journal of Archaeological Science* **3**, 275-277.

Kenward, H. K. (1974). Methods for Palaeoentomology on site and in the laboratory. *Science and Archaeology*, **15**, 16-24.

Kenward, H. K. (1990). The insect remains (1981). In, M. R. McCarthy, *A Roman, Anglian and Medieval Site at Blackfriars Street, Carlisle: Excavations 1977-9*, 317-318 & fiche 3/1-3. Cumberland & Westmoreland Antiquarian & Archaeological Society Research Series **4**. Carlisle.

Kenward, H. K. (1991). Buzz, buzz, buzz. A sceptical view of the Coppergate "beehive". *Interim: Archaeology in York* **15/4**, 20-24.

Kenward, H. K. & Allison, E. (1994) Rural origins of the urban and insect fauna. In A. R. Hall & H. K. Kenward (eds.) *Urban-Rural Connexions: Perspectives from Environmental Archaeology*, 55-78. Oxbow Monograph **47**, Oxford.

Kenward, H. K. & Hall, A. R. (1995) *Biological evidence from 16-22 Coppergate*. Archaeology of York, **14/7**. Council for British Archaeology for York Archaeological Trust, York.

Kenward, H. K., Hall, A. R., & Jones A. K. G. (1980). A tested set of techniques for the extraction of plant and animal macrofossils from waterlogged and archaeological deposits. *Science and Archaeology*, **22**, 3-15.

Kenward, H. K., Engleman, C., Robertson, A., Large, F. (1985). A tested set of techniques for the extraction of plant and animal macrofossils from waterlogged archaeological remains. *Circea*, **3**, 163-173.

Kenward, H. K. & Williams, D. (1979). Biological evidence from the Roman warehouses in Coney Street. *Archaeology of York*, **14/2**. Council for British Archaeology for York Archaeological Trust, London.

Killen, J. T. (1966). The Knossos Lc (Cloth) Tablets. *University of London, Institute of Classical Studies* **13**, 105-109.

Killen, J. T. (1985). The Linear B tablets and the Mycenean economy. In A. Morpugo-Davies and Y. Duhoux (eds.), *Linear B: A 1985 Survey. Proceedings of the Mycenaean Colloquium of the VIIIth Congress of the International Federation of the Societies of Classical Studies.* Louvain la Neuve, Cabay.

King, L. W. (1909). An early mention of cotton: the cultivation of *Gossypium arboreum* or tree-cotton in Assyria in the seventh century B.C. *Proceedings of the Society for Biblical Archaeology* **31**, 341.

Kirby, W. F. (1903). *The butterflies and moths of Europe*. London.

Kislev, M. E. (1991). Archaeobotany and storage entomology. In J. M. Renfrew (ed.) *New Light on Early Farming. Recent Developments in Palaeo-ethnobotany*, 121-136. Edinburgh University Press, Edinburgh.

Kislev, M. E., Artzy, M. & Marcus, E. (1993). Import of an Aegean food plant to a Middle Bronze Age IIA coastal site in Israel. *Levant* **25**, 145-154.

Knapp, A. B. (1991). Spice, drugs, grain and grog: organic goods in Bronze Age eastern Mediterranean trade. In N. H. Gale (ed.) *Bronze Age Trade in the Mediterranean. Studies in Mediterranean Archaeology* **90**, 21-68.

Knapp, A. B. & Cherry, J. F. (1994). *Provenience studies and Bronze Age Cyprus. Production, exchange and politico-economic change*. Monographs in World Archaeology **21**. Prehistory Press. Madison, Wisconsin.

Koch, K. (1971). Zur Untersuchung subfossiler Käferreste aus römer-zeitlichen und mittelalterlichen Ausgrabungen im Rheinland. *Rheinische Ausgrabungen* **10**, 378-448.

Koch, K. (1989) Ekologie **2**. *Die Käfer Mitteleuropas*, Goecke & Evers, Krefeld.

Koch, K. (1992) Ekologie **3**. *Die Käfer Mitteleuropas*, Goecke & Evers, Krefeld.

Kroll, H. J. (1982). Kulturpflazen von Tiryns. *Archäologischer Anzeiger* (1982), 467-485.

Kuniholm, P. I., Kromer, B., Manning, S. W., Newton, M., Latini, C. E., & Bruce, M. J. (1996). Anatolian tree rings and the absolute chronology of the eastern Mediterranean, 2220-718 BC. *Nature* **381**, 780-1.

Lambrou-Phillipson, C. (1990). Thera in the mythology of the Classical Tradition: An archaeological approach. In, D. A. Hardy, J. Keller, V. P. Galanopoulos, N. C. Flemming & T. H. Druitt *Thera and the Aegean World III*, **1**, 162-170. The Thera Foundation, London.

Latte, K. (ed.) (1953). *Hesychii Alexandrini Lexikon.* Ejnar Munksgaard, Copenhagen

Lesne, P. (1930). Le Dermeste des cadavres (*frischi* Kug.) dans les tombes de l' Égypte ancienne. *Bulletin de la Societé entomologique d' Égypte* (1930), 21-24.

Levinson, H. Z. & Levinson, A. R. (1985). Storage and insect species of stored grain in tombs in ancient Egypt. *Journal of Applied Entomology/ Zeitschrift fur angewandte Entomologie* **100**, 321-339.

Levinson, H. & Levinson, A. (1994). Origin of grain storage and insect species consuming desiccated food. *Anzieger Schädlingskunde, Pflanzenschutz, Umweltschutz* **67**, 47-59.

Lewis, C. D. (1940) *The Georgics of Vergil.* J. Cape, London.

Limbrey, S. (1990). Soil studies at Akrotiri. in D. A. Hardy, J. Keller, V. P. Galanopoulos, N. C. Flemming & T. H. Druitt (eds.) *Thera and the Aegean World III*, **2**, 377-383. The Thera Foundation, London.

Lindroth, C. H., Andersson, H., Bodvarsson, H. & Richter, S. H. (1973). *Surtsey, Iceland. The development of a new fauna, 1963-1970. Terrestrial invertebrates*. Entomologica Scandinavia, Suppl. **5**.

Liverani, M. (1962). La lettere del Faraone a Rib-Adda. *Oriens Antiquus* **10**, 253-268.

Liu, R. K. (1978). Spindle whorls Part I: Some comments and speculations. *The Bead Journal* **3**, 87-103.

Livy (1919-1959) *Livy in fourteen volumes.* (trans. by F. G. Moore, B. O. Foster, *et al.*) Loeb Text, Heinemann, London.

London, G., Egoumenidou, G. F. & Karageorghis, V. (1990). *Traditional pottery in Cyprus*. Mainz.

Lopez, R. S. (1945). Silk industry in the Byzantine Empire. *Speculum* **20**(1), 1-42.

Lubec, G. J., Holaubek C., Feldi B., Lubec, B. & Strouhal, E. (1993). Use of silk in Ancient Egypt. *Nature* **362**, 25.

Lucan (1928). *The civil war (Pharsalia)*, (trans. by J. D. Duff). Loeb Text, Heinemann, London.

Lucas, A. (1948). *Ancient Egyptian materials & industries* (3rd ed.). Edward Arnold & Co, London.

Luce, J. V. (1969). *The end of Atlantis*. Thames and Hudson, London.

Lucht, W. H. (1987). *Die Käfer Mitteleuropas. Katalog.* Goecke & Evers, Krefeld.

Mamet, H. (1874). *De insula Thera*. Lille.

Manning, S. W. (1990) The Thera eruption: the Third Congress and the problem of the date. *Archaeometry*, **32**, 91-100.

Manning, S. W. (1994). The emergence of divergence: Bronze Age Crete and the Cyclades. In C. Mathers & S. Stoddart (eds.) *Development and decline in the Bronze Age Mediterranean*, 221-270. Sheffield Archaeological Monographs. J. Collis Publications, Sheffield.

Manning, S. W. (1998). Correction. New GISP2 ice-core evidence supports 17th century BC date for the Santorini (Minoan) eruption: response to Zielinski & Germani (1998). *Journal of Archaeological Science* **25**, 1039-1042.

Marinatos, N. (1993). *Minoan religion: ritual, image and symbol.* University of South Carolina Press, Columbia.

Marinatos, S. (1939). The volcanic destruction of Minoan Crete. *Antiquity* **13**, 425-439.

Marinatos, S. (1976). *Excavations at Thera* **VII**. Athens.

Marinos, G. P. & Marinos, P. G. (1978). The ground water potential of Santorini island. In C. Doumas (ed.) *Thera and the Aegean World I*, 297-305. London.

Marshall, R. (1993). *Storm from the East.* BBC Publications, London.

Marthari, M. (1984). The destruction of the town at Akrotiri, Thera, at the beginning of LC I: Definition and chronology. In J. A. McGillivray and R. L. N. Barber (eds.) *The Prehistoric Cyclades*, 119-133.

Mauny, R. (1979). Contribution a l' étude des fosses ovoides et silos. In M. Gast & F. Sigaut (eds.) *Les techniques des conservation de grains a long terme*, I, 48-53. Centre National de la Recherche Scientifique. Paris.

Maxwell-Hyslop, K. R. (1970). An illustration to a Mari inventory? *Iraq* **32**, 165-166.

McCoy, F. W. (1990). The Upper Thera (Minoan) Ash in deep-sea sediments: distribution and comparison with other ash layers. In D. A. Hardy, J. Keller, V. P. Galanopoulos, N. C. Flemming & T. H. Druitt *Thera and the Aegean World III*, **2**, 57-78. The Thera Foundation, London.

McDermott, W. C. (1938). *The ape in Antiquity*. Baltimore.

Megas, G. E. (1968-9). Studies of popular architecture. *Laographia* **26**.

Meiggs, R. (1982). *Trees and timber in the Ancient Mediterranean World.* Clarendon Press, Oxford.

Melaart, J. (1962). Excavations at Çatal Hújuk. *Anatolian Studies* **13**, 43-103.

Mensina, J. & Renwick, J. A. A. (1983). Effectiveness of oils in protecting stored cowpeas from the cowpea weevil (Coleoptera: Bruchidae). *Journal of Economic Entomology* **76**, 634-636.

Merrillees, R. S. & Evans, J. (1980). An essay in provenance: the Late Minoan IB pottery from Egypt. *Berytus* **28**, 1-45.

Merrillees, R. S., & Winter, J. (1972). Bronze Age trade between the Aegean and Egypt: Minoan and Mycenaean pottery from Egypt in the Brooklyn Museum. *Miscellanea Wilbouriana* **I**, 101-133.

Michael, H. N. (1976). Radiocarbon dates from Akrotiri on Thera. In P. P. Betancourt (ed.) *Temple University Symposium* **I**, 7-9.

Michailidou, A. (1990). The lead weights from Akrotiri: The archaeological record. In D. A. Hardy, J. Keller, V. P. Galanopoulos, N. C. Flemming & T. H. Druitt *Thera and the Aegean World III*, **2**, 407-419. The Thera Foundation, London.

Miller, R. (1987). Appendix: ash as an insecticide. In B. J. Kemp, *Amarna Reports* IV, 14-16. Egypt Exploration Society, London.

Miller, W. (1904). (trans.) *Xenophon, Cyropaedia,* **II**. Heinemann, London.

Mills, J. O. (1992). Beyond nutrition: antibiotics produced through grain storage practices, their recognition and implications for the Egyptian Predynastic. In, R. Friedman & B. Adams (eds.) *The Followers of Horus. Studies dedicated to Michael Allen Hoffman*, 27-35. Egyptian Studies Association **2**, Oxbow Monograph **20**. Oxford.

Mills, J. W. & Gillespie, M. (1951). *Studio bronze casting.* Maclaren, London.

Minoranskiy, V. A. (1969). Some data on the ecology of the hide beetle *Dermestes laniarius* Illiger (Coleoptera, Dermestidae). *Entomological Review* **48**, 477-478.

Miracle, M. P. (1966). *Maize in Tropical Africa.* University of Wisconsin Press, Madison.

Monte, G. dal (1956). La presenza di insetti dei granai in frumento trovato negli scavi di Ercolano. *Redia* **41**, 23-28.

Morgan, L. (1987). *The miniature wall paintings of Thera. A study in Aegean culture and iconography.* Cambridge Classical Studies **19**. Cambridge University Press.

Morgan, L. (1995). Minoan painting and Egypt: The case of Tell el Dab'a. In W. M. Davies & L. Schofield (eds.) *Egypt, the Aegean and the Levant. Inter-connections in the Second Millennium BC*, 29-53.

Muhly, J. D. (1970). Homer and the Phoenicians. *Berytus* **19**, 19-64.

Mozley, J. H. (1979). (trans.) *Ovid, The Art of Love.* Loeb Text, Heinemann, London.

Munro, J. W. (1966). *Pests of stored products.* Hutchinson, London.

Muthesius, A. (1991). Crossing traditional boundaries: grub to glamour in Byzantine silk weaving. *Byzantine and Modern Greek Studies* **15**, 326-365.

Muthesius, A. (1993). The Byzantine silk industry: Lopez and beyond. *Journal of Medieval History* **19**, 1-67.

Nándor, I. (1966). Foldalatti gabonatarolas Magyarorszagon. *Ethnographia* **LXXVII**, 343-377.

Narrol, R. (1962). Floor area and settlement population. *American Antiquity* **27**, 587-589.

Needham, S. & Evans, J. (1987). Honey and dripping: Neolithic food residues from Runnymede Bridge. *Oxford Journal of Archaeology* **6**, 21-28.

Negbi, O. (1994). The "Libyan Landscape" from Thera: a review of Aegean enterprises overseas in the Late Minoan IA period. *Journal of Mediterranean Archaeology* 7, 101-140.

Nelson, D. E., Vogel, J. S., & Southon, J. R. (1990). Another suite of confusing radiocarbon dates for the destruction of Akrotiri. In D. A. Hardy, J. Keller, V. P. Galanopoulos, N. C. Flemming & T. H. Druitt (eds.) *Thera and the Aegean World III*, **3**, 197-206. The Thera Foundation, London.

Nicolaisen, J. (1963). *Ecology and culture of the pastoral Taureq*. Copenhagen National Museum.

Niemeier, W. D. (1990). Mycenaean elements in the miniature fresco from Thera? In D. A. Hardy (ed.) *Thera and the Aegean World III*, **1**, 267-284. The Thera Foundation. London.

Niemeier, W. D. (1991). Minoan artists travelling overseas: the Alalakh frescoes and the painted plaster floor at Tel Kabri (western Galilee). In R. Laffineur & L. Bach (eds.) *Thalassa. L' Egée Prehistorique et la Mer. Aegaeum* **7**, 189-201, Université de Liege.

Nikolaidis, N. J. (1959). [*Beekeeping: modern intensive methods*] 3rd ed. Athens: published by the author (in Greek).

Nilsson, M. P. (1950). *The Minoan-Mycenean religion and its survival in Greek religion*. G. W. K. Gleerup, Lund.

Noble, E. R. & Noble, G. A. (1976). *Parasitology*. Lea & Febiger, Philadelphia.

O' Farrell & Butler (1948). *Insects & mites in foodstuffs*. Economic Proceedings, Royal Dublin Society.

Oikonomides, N. (1986). Silk trade and production in Byzantium. *DOP* **40**, 33-53.

Oikonomou, M. (1992). [From the cocoon to silk cloth]. In Gagoulia *et al.* (eds) *Η σηροτροφία στο Σουφλί*. ETBA 72-107.

Ólafsson, E. & Richter, S. H. (1985). Húsamaurinn *Hypoponera punctatissima*. *Náttúrufræðingurinn* **55**, 139-146.

Oldroyd, H. (1964). *The natural history of flies*. Weidenfeld and Nicolson, London.

Oliva, A. (1939). I frumenti, le leguminose de granella e gli altri semi repertati a Belverde. *Studi Etruschi* **13**, 343-9.

Oppenheim, A. L. (1967). Essay on overland trade in the first millennium BC. *Journal of Cuneiform Studies* **21**, 236-54.

Osborne, P. J. (1969). An insect fauna of Late Bronze Age date from Wilsford, Wiltshire. *Journal of Animal Ecology* **38**, 555-566.

Osborne, P. J. (1971a). The insect fauna from the Roman harbour. In B. W. Cuncliffe, *Excavations at Fishbourne, 1961-1969*, 393-396. Research Reports of the Society of Antiquaries of London, **27**, Oxford.

Osborne, P. J. (1971b). An insect fauna from the Roman site at Alcester, Warwickshire. *Britannia* **2**, 156-165.

Osborne, P. J. (1974) An insect assemblage of Early Flandrian age from Lea Marston, Warwickshire and its bearing on the contemporary climate and ecology. *Quaternary Research* **4**, 471-486.

Osborne, P. J. (1977). Stored product beetles from a Roman site at Droitwich, England. *Journal of stored Product Research* **13**, 203-204.

Osborne, P. J. (1983). An insect fauna from a modern cesspit and its comparison with probable cesspit assemblages from archaeological sites. *Journal of Archaeological Science* **10**, 453-463.

Osborne, P. J. (1989) Insects. In P. Ashbee, M. Bell & E. Proudfoot *Wilsford Shaft. Excavations 1960-62*, 96-99 + fiche C1-7. English Heritage, London.

Osborne, P. J. (1997) Insects, man and climate in the British Holocene. In A. C. Ashworth, P. C. Buckland & J. P. Sadler (eds.) *Studies in Quaternary entomology - An inordinate fondness for insects. Quaternary Proceedings* **5**, 193-198.

Osella, G. (1974). Una nuova specie di *Troglorhynchus* Schmidt dell' Epiro. *Revue Suisse Zoologie* (Geneva) **81**, 791-795.

Osella, G. (1986). Due nuovi curculionidi endogei della fauna italiana (Coleoptera). *Entomol. Ital.* **33**, 373-383.

Osella, G. & Abbrazzi, P. (1985). Quattro nuove specie di Curculionidi dell' Apennino (Coleoptera). *Redia* **68**, 467-484.

Otten, H. (1967). Ein hethitiscer Vertrag aus dem 15./14. Jahrundert v. Chr. (KBo XVI 147). *IstMitt* **17**, 53-62.

Ovid (1947). *Heroides and Amores* (trans. by G. Showerman). Loeb Text, Heinemann, London.

Ovid (1979). *The Art of Love.* (trans. by J. H. Mozley). Loeb Text, Heinemann, London.

PAE (1992). Xeste 4. *Excavations at Santorini, Annual Report.*

Page, D. (1970). *The Santorini Volcano and the Desolation of Minoan Crete.* Supplement paper **12.** Society for the Promotion of Hellenic Studies, London.

Palladius, Rutilius Taurus Aemilianus (1773 & 1779). *On Husbandry.* (Lodge, B. & Herrtage, S. eds.). The Early English Text Society, **52** & **72.**

Palm, T. (1959). Die Holz und Rindenkäfer der sud- und mittelschwedischen Laubbaume. *Opuscula Entomologica Suppl.* **16.**

Pals, J. P. & Hakbijl, T. (1992). Weed and insect infestation of a grain cargo in a ship at the Roman fort of Laurium in Woerden (Province of Zuid-Holland). *Review of Palaeobotany and Palynology* **73**, 287-300.

Palyvou, C. (1984). The destruction of the town at Akrotiri, Thera, at the beginning of LC I: Rebuilding activities. In J. A. McGillivray & R. L. N. Barber (eds.) *The Prehistoric Cyclades,* 134-147.

Palyvou, C. (1986). Notes on the town plan of Late Cycladic Akrotiri, Thera. *British School of Archaeology* **81,** 179-194.

Palyvou, C. (1990). Architectural design at Late Cycladic Akrotiri. In, D. A. Hardy, J. Keller, V. P. Galanopoulos, N. C. Flemming & T. H. Druitt (eds) *Thera and the Aegean World* III, **1**, 44-56. The Thera Foundation, London.

Panagiotakopulu, E. (1989). *A palaeo-entomological study from the West House at Akrotiri, Santorini, Greece.* Unpubl. MSc diss., University of Sheffield.

Panagiotakopulu, E. (1999) An examination of biological materials from coprolites from XVIII Dynasty Amarna, Egypt. *Journal of Archaeological Science* **26**, 547-551.

Panagiotakopulu, E. & Buckland, P. C. (1991). Insect pests of stored products from Late Bronze Age Santorini, Greece. *Journal of stored Product Research,* **27**, 179-184.

Panagiotakopulu, E., Buckland, P. C., Day, P. M., Sarpaki, A., & Doumas, C. (1995). Natural insecticides and insect repellents in Antiquity: a review of the evidence. *Journal of Archaeological Science* **22**, 705-10.

Panagiotakopulu, E. & van der Veen, M. (1997) Synanthropic insect faunas from *Mons Claudianus*, a Roman quarry site in the Eastern Desert, Egypt. In A. C. Ashworth, P. C. Buckland & J. P. Sadler (eds.) *Studies in Quaternary Entomology - An Inordinate Fondness for Insects. Quaternary Proceedings* **5**, 199-206.

Pausanias (1918-35) *Pausanias' Description of Greece.* (trans. by V. H. S. Jones & H. A. Ormer) Loeb Text, Heinemann, London.

Parise, N. F. (1986). Unita ponderali egee. In M. Marazzi, S. Tusa and L. Vagnetti (eds.) *Traffici Micenei nel Mediterraneo*, 303-314, Tarentum.

Peacock, D. P. S. (1988). The Roman quarries of Mons Claudianus, Egypt. An interim report. In N. Herz & M. Waelkens (eds.) *Classical Marble: Geochemistry, Technology, Trade,* 97-102. Dordrecht.

Peacock, D. P. S & Maxfield, V. A. (1997) *Survey and excavation at Mons Claudianus 1987-1993.* **I.** *Topography and Quarries.* IFAO, Cairo.

Peacock, E. R. (1993). *Adults and larvae of hide, larder and carpet beetles and their relatives (Coleoptera: Dermestidae) and of derodontid beetles (Coleoptera: Derodontidae).* Handbooks for the identification of British insects **5**, 3.

Pelekassis, C. E. D. (1962). A catalogue of the more important insects and other animals harmful to the agricultural crops of Greece during the last thirty-year period. *Annals of the Institute of Phytopathology* (Benaki) N.S. **5.**

Pereira, J. (1983). The effectiveness of six vegetable oils as protectants of cowpeas and bambara groundnuts against infestation by *Callosobruchus maculatus* (F.) (Coleoptera: Bruchidae). *Journal of stored Product Research* **19**, 57-62.

Perrot, G. & Chipiez, C. (1894). *History of the Arts in Primitive Greece*, (2 Vols). Chapman and Hall, London.

Perrot, J. (1957). Le Mesolithique de Palestine et les recentes decouvertes a Eynan (Ain Mallaha). *Antiquity and Survival* **2**, 91-110.

Perrot, J. (1960). Excavations at Eynan (Ein Mallaha), Preliminary Report on the 1959 Season. *Israel Exploration Journal* **10**,14-22.

Petruso, K. M. (1978). Marks on some Minoan balance weights and their interpretation. *Kadmos* **17**, 26-42.

Pichler, H. & Schiering, W. (1977). The Theran eruption and Late Minoan IB Destruction on Crete. *Nature* **267**, 819-22.

Piggott, S. (1958). Native Economies and the Roman Occupation of North Britain. In I. A. Richmond (ed.) *Roman and Native in North Britain*, 1-27. Nelson, London.

Plato (1929). *Timaeus; Critias; Creitophon; Menexenus; Epistles* (trans. by R. G. Bury). Loeb Text, Heinemann, London.

Plautus (1916). *Curculio.* (trans. by Paul Nixon). Loeb Text, Heinemann, London.

Pliny (1938-63). *Natural History,* (trans. H. Rackham) (8 vols.). Loeb Text, Heinemann, London.

Plutarchus (1976). *Moralia,* XIII (trans. by Harold Cherniss). Loeb Text, Heinemann, London.

Polunin, O. & Huxley, A. (1965). *Flowers of the Mediterranean.* Chatto & Windus, London.

Postgate, J. P. (1913). (trans.) *Tibullus, Catullus, Tibullus and Pervigilium Veneris.* Loeb Text, Heinemann, London.

Potts, D. T. (1992). *The Arabian Gulf in Antiquity,* **II**. Clarendon Press, Oxford.

Poulianos, A. N. (1972). The discovery of the first victim of Thera's Bronze Age eruption. *Archaeology* **25**, 229-230.

Procopius (1935). *The Anecdota* or *Secret History* (trans. by H. B. Dewing). Loeb Text, Heinemann, London.

Procopius (1905). *De Bello Gothico.* (trans. by J. Havry). Leipzig.

Propertius, Sextus (1912) *Propertius.* (trans. by H. E. Butler). Loeb Text, Heinemann, London.

Pyle, D. M. (1989). Ice core acidity peaks, retarded tree growth and putative eruptions. *Archaeometry* **31**, 88-91.

Pyle, D. M. (1990). The application of tree-ring and ice core studies to the dating of the Minoan eruption. In D. A. Hardy, J. Keller, V. P. Galanopoulos, N. C. Flemming & T. H. Druitt (eds.) *Thera and the Aegean World* III, **1**, 113-121. The Thera Foundation, London.

Rackham, H. (1940-63) (trans.) Pliny, *Natural History.* Loeb Text, Heinemann, London.

Rackham, O. (1978). The flora and vegetation of Thera before and after the great eruption. In C. Doumas (ed.), *Thera and the Aegean World* I, 755-754. The Thera Foundation, London.

Rackham, O. (1982). Land-use and the native vegetation of Greece. In M. Bell & S. Limbrey (eds.) *Archaeological Aspects of Woodland Ecology*, 177-196. British Archaeological Reports International Series **146**, Oxford.

Rackham, O. (1990). Observations on the historical ecology of Santorini. In D. A. Hardy (ed.) *Thera and the Aegean World* III, **2**, 384-391. The Thera Foundation, London.

Rahmann, A. K. (1944). *Indian Farming* **5**, 272-275.

Rau, D. (with an introduction by Hale, A. & FitzGibbon, K.) (1988). *Ikats woven silks from Central Asia The Rau Collection.* Basil Blackwell, Oxford.

Reavey, D., Archer, M., Crossley, R., Douglas, A., Kenward, H., Port, G., & Gaskin, J. (organising commitee) (1994). *5th European Congress of Entomology 29th August-2 September 1994. Abstracts.* University of York.

Redding, R.W. (1991). The role of the pig in the subsistence system of ancient Egypt. In P. Crabtree and K. Ryan (eds.) *Animal Use and Culture Change*, 21-30. MASCA, The University Museum of Archaeology and Anthropology, University of Pennsylvania, Philadelphia.

Regnault, C., Hamraoui, R. & Hamraoui, A. (1993). Efficiency of plants from the South of France used as traditional protectants of *Phaseolus vulgaris* L. against its bruchid *Acanthoscelides obtectus*. *Journal of stored Product Research* **29**, 3, 259-264.

Renfrew, A. C. & Wagstaff, M. (eds.) (1982). *An island polity: The archaeology of exploitation in Melos*. Cambridge University Press.

Renfrew, J. (1985) Botanical report. In B. J. Kemp *Amarna Reports* **II**, 182. Egypt Exploration Society, London.

Reut, M. (1983). *La soie en Afganistan. L' elevage du ver à soie en Afganistan et l' artisanat de la soie à Herat*. Wiesbaden.

Reynolds, P. J. (1974). Experimental Iron Age storage pits: an interim report. *Proceedings of the Prehistoric Society* **40**, 118-131.

Reynolds, P. J. (1979). A general report of underground grain storage experiments at the Butser Ancient Farm research project. in M. Gast & F. Sigaut (eds.) *Les Techniques des Conservation de Grains à Long Terme* I, 70-80. Centre National de la Recherche Scientifique. Paris.

Reynolds, V (1968). *The apes: the gorilla, chimpanzee, orangutan, and gibbon: their history and distribution*. Cassel, London.

Richards, O. W. (1974). The Cretan "hornet" pendant. *Antiquity* **48**, 222.

Richards, O. W. (1975). The Cretan "hornet". *Antiquity* **49**, 212-213.

Richter, G. M. A. (1929). Silk in Greece. *American Journal of Archaeology* **33**, 27-33.

Rickman, G. (1971). *Roman granaries and store buildings*. Cambridge University Press, Cambridge.

Roberts, M. J. (1985). *The spiders of Great Britain and Ireland*. E. J. Brill, Leiden.

Robinson, M. A. (1979) The biological evidence. In G. Lambrick & M. A. Robinson, *Iron Age and Roman Riverside Settlements at Farmoor, Oxfordshire*, 77-133. Council for British Archaeology Research Report, **32**, London.

Robinson, M. A. (1991). *The Neolithic and Late Bronze Age insect assemblages*. In S. Needham, *Excavation and Salvage at Runnymede Bridge, 1978: The Late Bronze Age Waterfront Site*, 277-326. British Museum Press, London.

Robinson, S. (1969). *A history of dyed textiles*. Cambridge, Mass.

Rogers, B. B. (1924) (trans.) *Aristophanes Lysistrata*. Loeb Text, Heinemann, London.

Roper, T. (1996) Fossil insect evidence for the development of raised mire at Thorne Moors, near Doncaster. *Biodiversity and Conservation*, **5**, 503-521.

Rosetti, G. V. (1540). *Plichto de Latre de Tentori*. Venice.

Rothschild, M. & Clay, T. (1952). *Fleas, flukes and cuckoos: a study of bird parasites*. Collins, London.

Rowley-Conwy, P. & Zvelebil, M. (1989). Saving it for later: storage by prehistoric hunter-gatherers in Europe. In P. Halstead & J. O' Shea (eds.) *Bad year economics: cultural responses to risk and uncertainty*, 40-56. Cambridge University Press, Cambridge.

Ruttner, F. (1979). Ein Bienenkorb von der Nordseeküste aus prähistorischer Zeit. *Bienenmuseum und Geschichte der Bienenzucht. Internationales Symposium über Bienenwirtschaft (Freiburg), 16-18 August 1977* (Bucharest), 109-119.

Ryder, M. (1995). *Archaeological Textiles Newsletter* **20**, 26.

Sadler, J. P. (1991) *Archaeological and palaeoecological implications of palaeoentomological studies in Orkney and Iceland*. Unpubl. PhD thesis University of Sheffield.

Saggs, H. W. F. (1989). *Civilization before Greece and Rome*. Batsford, London.

Sahlins, M. (1972). *Stone Age economics*. Aldine-Atherton, Chicago & New York.

Saim, N. & Meloan, C. E. (1986). Compounds from leaves of bay (*Laurus nobilis* L.) as repellants for *Tribolium castaneum* (Herbst) when added to wheat flour. *Journal of stored Product Research* **22**, 141-144.

Sakellarakis, J. A. (1990). The fashioning of ostrich-egg rhyta in the Creto-Mycenean Aegean. In D. A. Hardy (ed.) *Thera and the Aegean World III*, **1**, 285-308. The Thera Foundation, London.

Sarpaki, A. A. (1987). *The palaeoethnobotany of the West House, Akrotiri, Thera. A case study*. Unpubl. PhD thesis, University of Sheffield.

Sarpaki, A. A. (1990). Small fields or big fields? That is the question. In C. Doumas (ed.) *Thera and the Aegean World III* **2**, 422-432. The Thera Foundation, London.

Schaeffer, C. F. A. (1933). Les fouilles de Minet el-Beidha et de Ras Shamra. *Syria* **14**, 93-117.

Schaeffer, C. F. A. (1939). *Ugaritica I. Mission de Ras Shamra* **3**. Bibliotheque Archeologique et Historique **47**. Geuthner: Paris.

Schelvis, J. (1990) Mites (Acari) in archaeology. *Experimental and Applied Entomology, Proceedings of the Netherlands Entomological Society*, **1**, 90-95.

Schelvis, J. (1992). *Mites and archaeozoology. General methods; applications to Dutch sites*. Ph.D thesis, University of Groningen.

Schelvis, J. (1997) Mites in the background: Use and origin of remains of mites (Acari) in Quaternary deposits. In, A. C. Ashworth, P. C. Buckland & J. P. Sadler (eds.) *Studies in Quaternary entomology - An inordinate fondness for insects. Quaternary Proceedings* **5**, 233-236.

Schoonhoven, A. V. (1978). Use of vegetable oils to protect stored beans from bruchid attack. *Journal of economic Entomology* **71**, 246-254.

Scott, P. (1993). *The book of silk*. Thames & Hudson, London.

Seitz, A. (1913). *The Macrolepidoptera of the World*. Stuttgart.

Sextus Empiricus (1933). *Outlines on Pyrrhonism* I. (trans. R. G. Bury). Loeb Text, Heinemann, London.

Shaw, I. (1984). Report on the 1983 excavations. The animal pens (Building 400). In B. J. Kemp (ed.), *Amarna Reports* **I**, 40-56. Egypt Exploration Society, London.

Shaw, J. W. (1990). Bronze Age Aegean harboursides. In D. A. Hardy, J. Keller, V. P. Galanopoulos, N. C. Flemming & T. H. Druitt *Thera and the Aegean World III*, **1**, 420-436. The Thera Foundation, London.

Shelmerdine, C. W. (1985). *The perfume industry of Mycenean Pylos* (Studies in Mediterranean Archaeology pocket-book, **34**). Åstrom, Goteborg.

Sherratt, A. G. & Sherratt, E. S. (1991). From luxuries to commodities: the nature of Mediterranean Bronze Age trading systems. In N. H. Gale (ed.) *Bronze Age Trade in the Mediterranean. Studies in Mediterranean Archaeology* **90**, 351-356.

Shewring, W. (1980) (trans.) *Homer. The Odyssey* Oxford University Press, Oxford.

Sigaut, F. (1978). *Lés reserves de grains a long terme*. Editions de la Maison des Sciences de l' Homme, University of Lille.

Sigaut, F. (1988). A method for identifying grain storage techniques and its application for European agricultural history. *Tools and Tillage* **6**, 3-32.

Sighamony, S., Anees, I., Chandrakala, T. & Osmani, Z. (1986). Efficacy of certain indigenous plant products against *Sitophilus oryzae* (L.) and *Rhyzopertha dominica* (F.). *Journal of stored Product Research* **22**, 21-23.

Skidmore, P. (1996). *A dipterological perspective on the Holocene history of the North Atlantic area.* Unpubl. Ph.D., University of Sheffield.

Smith, J. & Ross, W. (1910) (trans.) Aristotle. *Historia Animalium.* Oxford University Press, Oxford.

Solomon, M. E. (1965). Archaeological records of storage pests: *Sitophilus granarius* (L.) (Coleoptera, Curculionidae) from an Egyptian pyramid tomb. *Journal of Stored Product Research* **1**, 105-107.

Solomon, M. E. & Adamson, B. E. (1955). The powers of survival of storage and domestic pests under winter conditions in Britain. *Bulletin of Entomological Research* **46**, 311-355.

Sotirakopoulou, P. (1986). Early Cycladic pottery from Akrotiri on Thera and its chronological implications. *Annual Report of the British School in Athens* **81**, 297-312.

Southgate, B. J. (1978). The importance of the Bruchidae as pests of grain legumes, their distribution and control. In S. R. Singh, H. F. van Emden and T. A. Taylor (eds.) *Pests of Grain Legumes: Ecology and Control.* London, Academic Press.

Sparks, R. S. J. & Wilson, C. J. N. (1990). The Minoan deposits: A review of their characteristics and interpretation. in D. A. Hardy, J. Keller, V. P. Galanopoulos, N. C. Flemming & T. H. Druitt (ed.) *Thera and the Aegean World III,* **2**, 89-99. The Thera Foundation, London.

Spencer, A. J. (1982). *Death in Ancient Egypt.* Penguin, Harmondsworth.

Stork, L. (1975). Antilope, *Lexikon der Agyptologie*. Wiesbaden **I**, 319-323.

Strabo (1930). *The Geography of Strabo*, 7, (trans. by H. L. Jones). Loeb Text, Heinemann, London.

Strange, J. (1980). *Caphtor/Keftiu. Acta Theologica Danica* **14**. Brill, Leiden.

Strobel, P. & Pigorini, L. (1864). Le terremare e le palafitte del Parmense, seconda relazione. *Atti della Societa italiana di Scienze Naturali* (Milan), 7, 36-37.

Strong, L. (1981). Dermestids - an embalmer's dilemma. *Antenna* **5**, 136-139.

Strouhal, E. (1996) *Life of the Ancient Egyptians*. Liverpool University Press, Liverpool.

Sumner, W. (1979). Estimating population by analogy: an example. In C. Kramer (ed.) *Ethnoarchaeology: Implications of Ethnography for Archaeology*, 164-174. Columbia University Press. New York.

Sundwall, J. (1932a). *Zu dem minoischen Wahrungssystem*. Mélanges Glotz II.

Sutton, M. Q. Archaeological aspects of insect use. *Journal of Archaeological Method and Theory* **2**, 253-297.

Tertullianus, Quintus Sept. Flor. (1853). *Quae supersunt Omnia*, I. *-De Pallio*. T. O. Weigel.

Tertullianus, Quintus Sept. Flor. (1895). *The writings, III. -On the Ascetic's Mantle. With the extant works of Victorinus and Commodianus*. Edinburgh.

Theophrastus (1916). *Historia Plantarum*, (trans. by A. Hort) (2 vols.). Loeb Text, Heinemann, London.

Theophrastus (1990). *De causis plantarum* (trans. B.Einarson & G. K. K. Link) (2 vols.). Loeb Text, Heinemann, London.

Thucydides (1959). *The Peloponnesian War*, (trans. by T. Hobbes). Ann Arbor, Michigan University Press.

Tibullus (1913). *Catullus, Tibullus and Pervigilium Veneris* (trans. J. P. Postgate). Loeb Text, Heinemann, London.

Trantalidou, C. (1990). Animals and Human Diet in the Prehistoric Aegean. In D. A. Hardy, J. Keller, V. P. Galanopoulos, N. C. Flemming & T. H. Druitt (eds.) *Thera and the Aegean World III*, **2**, 392-405. The Thera Foundation, London.

Trinatafyllidou-Baladie, Y. (1979). Greniers publics et familiaux en Grece du XVIeme au XXeme siecle. In M. Gast & F. Sigaut (eds.) *Les techniques des Conservation de Grains a Long Terme*, I, 150-159. Centre National de la Recherche Scientifique. Paris.

Trinatafyllidou-Baladie, Y. (1988). *Το εμπόριο και η οικονομία της Κρήτης από τις αρχές της Οθωμανικής Κυριαρχίας έως το τέλος του 18ου αιώνα (1669-1795)*, 204-297. Ηράκλειον Κρήτης.

Tyagi, A. K. & Girish, G. K. (1975). Studies in the assessment of storage losses of food grains by insects. *Bulletin of Grain Technology* **13**, 84-102.

Tyers, I. (1988) Environmental evidence from Southwark and Lambeth. In P. Hinton (ed.) *Excavations in Southwark 1973-76 Lambeth 1973-79*, 443-477. Jnt Publ **3**, London & Middlesex Archaeological Society & Surrey Archaeological Society. London.

Tzachili, I. (1990). All important yet elusive: looking for evidence of cloth-making at Akrotiri, in D. A. Hardy, J. Keller, V. P. Galanopoulos, N. C. Flemming & T. H. Druitt (eds.) *Thera and the Aegean World III*, **1**, 380-390. The Thera Foundation, London.

Tzachili, I. (1992). Small finds from the Akrotiri excavation. 'Ακρωτήρι Θήρας, Εἴκοσι χρόνια ἔρευνας (1967-1987). Ἡμερίδα 'Αθῆναι, (19 Dec. 1987), 139-146.

Valamoti, S. M. & Buckland, P. C. (1995). An early find of *Oryzaephilus surinamensis* L. (Coleoptera: Silvanidae) from Final Neolithic Mandalo, Macedonia, Greece. *Journal of stored Product Research* **31**, 307-309.

van de Bund, C. F. (1988). Insecten en mijtenresten uit een 13e eeuwse kuil met organisch materiaal in 's-Hertogenbosch. In H. W. Boekwijt & H. L. Janssen (eds.), *Kroniek von bouwhistorisch en archaeologisch onderzoek's Hertogen-bosch*, 150-153.

Vandersleyen, C. (1967). Une tempete sous le regne d'Amosis. *Revue d' Egyptologie* **19**, 123-159.

van Emden, F. I. (1951) On the genus *Thaumphrastus* Blaisdell (Coleoptera: Thorictidae). *Bulletin of the Brooklyn Entomological Society*, **46**, 39-41.

Vanschoonwinkel, J. (1990). Animal re-presentations in Theran and other Aegean arts, in D. A. Hardy (ed.) *Thera and the Aegean World III*, **1**, 327-349. The Thera Foundation, London.

Varro, Terentius (1967). *On Agriculture* (trans. by Hooper W. D.). Loeb Text, Heinemann, London.

Vartavan, C. de (1990). Contaminated plant-food from the tomb of Tutankhamun:

a new interpretive system. *Journal of Archaeological Science* **17**, 473-494.

Veen, M. van der (1996). The plant remains from Mons Claudianus, a Roman quarry settlement in the Eastern Desert of Egypt - an interim report. *Vegetation History and Archaeobotany* **5**, 137-141.

Veen, M. van der (1998) A life of luxury in the desert? The food and fodder supply to Mons Claudianus. *Journal of Roman Archaeology* **11**, 101-116.

Venitzelos, E. (1990). New palaeobotanical data for the evolutionary history of plants in the Aegean area, with special reference to the palaeoflora of Thera. In D. A. Hardy, J. Keller, V. P. Galanopoulos, N. C. Flemming & T. H. Druitt (eds.) *Thera and the Aegean World III,* **2**, 406-415. The Thera Foundation, London.

Ventris, M. & Chadwick, J. (1973). *Documents in Mycenaean Greek.* (2nd ed. J. Chadwick) Cambridge University Press, Cambridge.

Vercoutter, J. (1956). *L' Égypte et le monde egeen prehellenique.* L' Institut Francais d' Archaeologie Orientale, Cairo.

Vergil (1871). *Georgics.* (trans. by J. Lonsdale & L. S. Macmillan). London.

Verneau R. (1891). *Cinq années de séjour aux îles Canaries.* Hennuyer, Paris.

Virgil (1990). *Georgics*, (R. A. B. Mynors, ed.). Clarendon Press, Oxford.

Vitruvius (1960). *On architecture*, (trans. by M. H. Morgan). New York.

Wagstaff, J. M., Augustson, S. & Gamble, C. (1982). Alternative subsistence strategies. In C. Renfrew & J. M. Wagstaff (eds.). *An Island Polity*, 172-180. Cambridge University Press.

Wagstaff, J. M., & Cherry, J. F. (1982). Settlement and population change. In C. Renfrew and J. M. Wagstaff (eds.). *An Island Polity*, 136-155. Cambridge University Press.

Wallis Budge, E. A. (1925). *The Mummy. A handbook of Egyptian funerary archaeology.* (2nd ed.). Routledge, Kegan & Paul, London.

Waloff, Z. (1946). *Seasonal breeding and migration of the desert locust in Eastern Africa.* Anti-Locust Research Centre, BMC, London.

Wardle, K. A. (1987). Excavations at Assiros Toumba 1986. A preliminary report. *Annual of the British School of Archaeology at Athens* **82**, 313-329.

Wardle, T. (1891). On the history of growing utilisations of tussur silk. *Journal of the Society of Arts* (1891), 3-47.

Wardle, T. (1904). *Kashmir, its new silk industry.* Simpkin, Marshall, Hamilton, Kent & Co. London.

Warren, P. (1987). The Ring of Minos, in Είλαπινή. Τόμος Τιμητικός για τον Καθ. Ν. Πλάτωνα, 485-500. Herakleion.

Warren, P. (1990a). Summary of evidence for the absolute chronology of the early part of the Aegean Late Bronze Age derived from historical Egyptian sources, in D. A. Hardy (ed.) *Thera and the Aegean World III*, **3**, 24-6. The Thera Foundation, London.

Warren, P. (1990b). Of Baetyls. *Opuscula Atheniensia* XVIII, 193-206.

Warren, P. (1995). Minoan Crete and Pharaonic Egypt. In W. V. Davies & L. Schofield. (eds.). *Egypt, the Aegean and the Levant.* Interconnections in the Second Millennium BC, 1-18. British Museum Press, London.

Weinstein, G. A. & Betancourt, P. P. (1978). Problems of interpretation of the Akrotiri radiocarbon dates. In C. Doumas (ed.), *Thera and the Aegean World* I, 805-814. The Thera Foundation, London.

Wertime, T. A. (1983). The furnace versus the goat: the pyrotechnologic industries and Mediterranean deforestation in antiquity. *Journal of Field Archaeology* **10**, 445-52.

Wheeler, G. (1682). *A journey into Greece.* T. Cademan, London.

White, H. G. E. (1929) (trans.). Hesiod *The homeric hymns and homerica.* Loeb Text, Heinemann, London.

White, K. D. (1970) *Roman Farming.* Thames & Hudson, London.

White, R. E. (1990). *Lasioderma haemorrhoidale* (Ill.) now established in California, with biological data on *Lasioderma* species. *Coleopterists Bulletin* **44**, 344-348.

Wild, J. P. (1984). Some early silk finds in Northwest Europe. *Textile Museum Journal* **23**, 17-23.

Wilkinson, R. H. (1992). *Reading Egyptian Art. A hieroglyphic guide to ancient Egyptian painting and sculpture.* Thames & Hudson, London.

Woodroffe, G. E. (1953). An ecological study of the insects and mites in the nests of certain birds in Britain. *Bulletin of Entomological Research* **44**, 739-772.

Woodroffe, G. E. (1967). Phoretic behaviour of adult *Aglenus brunneus* (Gyllenhal) (Col., Colydiidae). *Entomologist's Monthly Magazine* **103**, 44.

Wrigley, C. (1960). Speculations on the economic prehistory of Africa. *Journal of African History* **1**, 189-203.

Xenophon (1904). *Cyropaedia,* **II**, (trans. by W. Miller). Heinemann, London.

Yannai, A. (1983). *Studies on trade between the Levant and the Aegean in the 14th to 12th centuries B. C.* Unpubl. D. Phil. dissertation. University of Oxford.

Younger, J. P. (1983). Aegean seals of the Late Bronze Age. Masters and workshops III: The first generation Minoan masters. *Kadmos* **22**, 109-36.

Yvinec, J.-H. (1997) Infestation par les insectes d'un grenier carbonisé de la fin du IIe siècle A.D. à Amiens (Somme). *L'Entomologiste* (Paris) **53**, 113-128.

Zacher, F. (1927). *Die Vorrats-, Speicher- und Materialschadlinge und ihre Bekämpfung.* Berlin.

Zacher, F. (1934). Vorratsschadlinge und Speicherwirtschaft im alten und neuen Ägypten. *Forschung und Fortschritte* **10**, 347-348.

Zacher, F. (1937). Vorratschädlinge und Vorratschutz, ihre Bedeutung für Volksernahrung und Weltwirtschaft. Zeitschrift für hygienische *Zoologie und Schädlingsekämpfung* **29**, 1-11.

Zander, E. (1941). *Beitrage zur Herkunftsbestimmung bei Honig. III. Pollengestaltung und Herkunftsbestimmung bei Blutenhonig (Erganzungsband).* Liedloff, Loth & Michaelis, Leipzig.

Zannetti, T. (1980). *A field guide to insects in colour.* Octopus Books, London.

Zeuner, F. E. (1963). *A history of domesticated animals.* Hutchinson and Co., London.

Zielinski, G. A. & Germani, M. S. (1998a). New ice-core evidence challenges tohe 1620s BC age for the Santorini (Minoan) eruption. *Journal of Archaeological Science* **25**, 279-289.

Zielinski, G. A. & Germani, M. S. (1998b). Reply to: Correction. New GISP2 ice-core evidence supports 17th century BC date for the Santorini (Minoan) eruption. *Journal of Archaeological Science* **25**, 1043-1045.

Zielinski, G. A., Mayewski, P. A., Meeker, L. D., Whitlow, S., Twickler, M. S., Morrison, M., Meese, D. A., Gow, A. J., & Alley, R. B. (1994). Record of volcanism since 7000 BC from the GISP2 Greenland Ice Core and Implications for the Volcano-Climate System. *Science* **264**, 948-952.

Zisis, V. G. (1955). Cotton, linen and hempen textiles from the Fifth century BC. *Πρακτικά τῆς Ἀκαδημίας Ἀθηνῶν* **29**, 587-93.

Zumpt, F. (1965). *Myiasis in Man and Animals in the Old World.* Butterworth & Co. London.

APPENDIX 1

REFERENCES IN THE CLASSICAL RECORD TO MEASURES AGAINST PEST INFESTATION

1. Proper time - weather - climate - regions

Pliny	HN XVIII LXXII 303
Geoponika	II 25
Columella	RR II 20.5 (weather)
Theophrastus	HP VIII 11 3 (climate)
	CP IV 2.2 (areas)

2. Mechanical control

Theophrastus HP VIII 10 2	Achillean barley-barley to wheat
CP IV 16.2, 2.2	nature of seed
Pliny HN XIX LVII	
HN XIX LVIII 181	longevity in regard to the deterioration
HN XI XXXV 105-106	control of locusts

3. Before sowing

Pliny HN XVIII XLV 160	houseleek deep (Demokritus)
HN XVIII XLV 158	olive lees+hoeing
	wine+water cypress leaves
Theophrastus CP III 24.4	wine
Virgil Georg. I 193-4	nitre+black oil lees

4. In the field

Pliny HN XVII XXXVIII 234	pitch-oil-grease: young trees olive lees+dung: fig trees
HN XIX LVII 178	dung on the leeks
HN XIX LVIII 179	wormwood or houseleek for cabbages
HN XVIII XLV 158	ashes
Cato XCV	amurca+bitumen+sulphur around the trank and branches of vines
Columella RR XIV	lupines+bitumen or oil+bitumen on vines against ants
Pliny HN XVIII LXX 294	smoke against blight
Aristotle HA iv.8	burning storax
Pliny HN XIX LVII 180	burning galbanum resin against gnats

5. Threshing floor

Pliny HN XVIII LX	amurca
Cato DA CXXIX	amurca
Virgil Georg I 178	cowdung solution

6. Construction of buildings

Vitruvius	A VI 6 4
Varro	RRm I LVII.1
Columella	RR I VI.10
	RR I XVI.12
Pliny	HN XVIII LXXII 301-2

7. Interiors of storerooms

Varro	RRm LVII 1
Cato	DA XCII
Columella	RR I VI.13-5
Pliny	HN XVIII XXXVI 136

8. Underground storage

Cato	DA XCII
Varro	RRm I LVII 2
Columella	RR I VI.15
Pliny	HN XVIII LXXIII 306-7

9. During storage

Theophrastus CP IV 6.5	winnowing
Columella RR I VI.17	damaged grain undisturbed
Columella RR II X.16	"
Varro RRm I LVII 2	"
Palladius I.19.3	"
Pliny HN XVIII 302	"
Pliny HN XVIII LXXXIII 305	olive-oil lees wormwood
	Carian-Chalcic chalk
	Olynthian Cerinthian earth
Pliny HN XVIII LXXIII 308	lentils treated with vinegar and silphium
	salted fish
Columella RR II.X.16	oil jars-ashes
Varro RRm I LVIII 1	
Theophrastus CP IV 13.7	earth thrown on the grain
Palladius	coriander-fleabane foliage
Plutarch Quast. Conv. V.5.1.	clay with grain
	grapes+hard pitch with dry chaff
Pliny HN XIX LVIII 178	seaslimes ashes
	heliotrope
	against ants water in
	which a brick was put in
Theophrastus CP VI 5.3	smell of olive oil
CP VI 5.4	marjoram
Cato DA CLXII	for preserving ham: salt and
	oil and vinegar keep the
	moths and worms away
Aristotle HA	oil kills insects
DMA 845b	smell of roses kills
Theophrastus CP VI 5.1	antipathy to roses
Aelian DNA 1.38	beetles die of smell of roses
Aelian DNA 6.46	insects die of oil
Aelian DNA 4.18	beetle dies if you put it on roses
Plutarch Moral. 710E	antipathy to perfume
1058A	
Plutarch Moral. 1096A	antipathy to perfume and incense
Sextus Empiricus Pyrrh I.55	perfume

Theophrastus O 4	
Dio Chrysostom orat. 32.98	
Hipp. Cant. LXXI-13	oil
Hipp. Berol LXXXV 4-7	oil-ashes against parasites of horses
Aristotle HA iv.8 (534b 48-25)	sulphur against ants: sprinkling ground marjoram and sulphur

APPENDIX 2

THE PAPYRUS EBERS; XCVII-XCVIII

"The beginning of remedies prepared to expel fleas in a house: thou shalt besprinkle it with natron-water, until (they) pass away.

Another: *bbt*-plant, is ground with charcoal, the house is coated therewith thoroughly, until (they) pass away.

Another to prevent a serpent from coming out of its hole: a dry *int*-fish, placed in the opening of its hole; it will not come out therefrom.

Another: natron, placed in the opening of the hole; it will not come out. Another: a bulb of onion, placed in the opening of the hole; it will not come out.

Another to prevent fly from biting: fat of *gnw*-bird; (it) is anointed therewith.

XCVIII Another to prevent gnats from stinging: fresh balanites-oil (it) is rubbed therewith.

Another to prevent mice from approaching to things: fat of cat, is placed on all things.

Another to prevent the kite from robbing: a branch of acacia, let it stand up. The man shall say: oh, Horus, he has stolen in town and in the field; his thirst is for the birds' field; he shall be cooked and eaten. Is recited over a branch of acacia, and *fk3*-cakes are applied to it; that is the way to prevent the kite from robbing.

Another to prevent *kkt*-animals from eating corn in the barn: excrements of gazelle, are placed over a fire in the barn, and its walls and its earthen floor are coated with their dirt and water; that is the way to prevent the corn being eaten.

Another, what is done to kill *smr*-animals, are placed over the fire until it dies.

Incense, what is done to sweeten the smell of the house or the clothes: dry myrrh, pingon, frankincence, rush-nut, wood (i.e. bark) of cinnamon, sbt, reed from Phoenicia (*Calamus aromaticus*), *inkwwn*, *dmtn*, liquid styrax, are ground fine, mixed together and (a little) thereof is placed over a fire.

Another (use) made by women therewith: to put this remedy, (made) according to this direction, into honey, is boiled, mixed and shaped into balls; they shall fumigate with them. But they use a taste of them to sweeten the smell of their mouth".

(Ebell 1937)

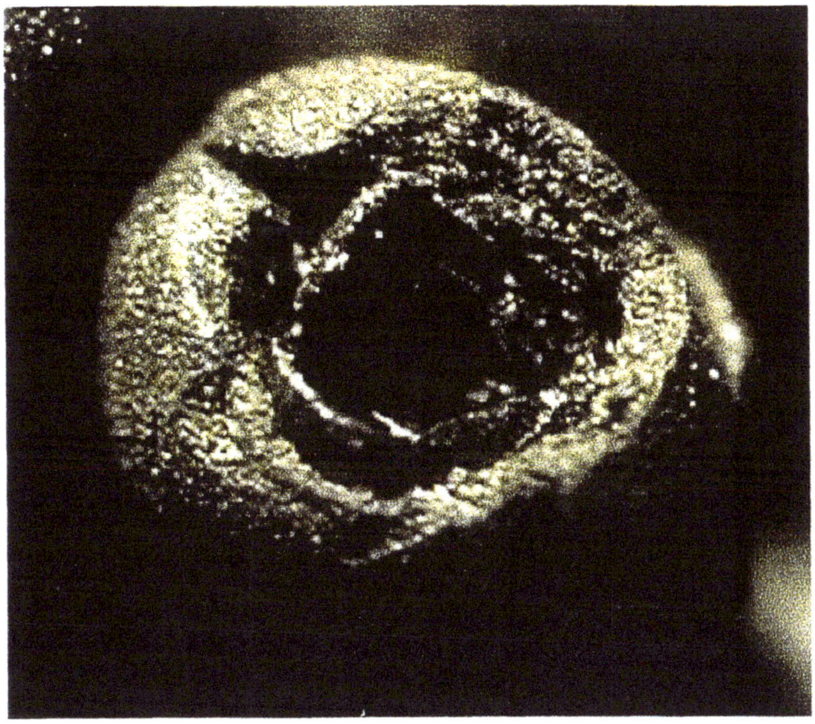

Pl. 1. Seed of *Lathyrus clymenum* from Akrotiri, Thera, Greece, charred during the eruption. An adult bruchid beetle can be seen emerging from the hole. (L. 3mm.)

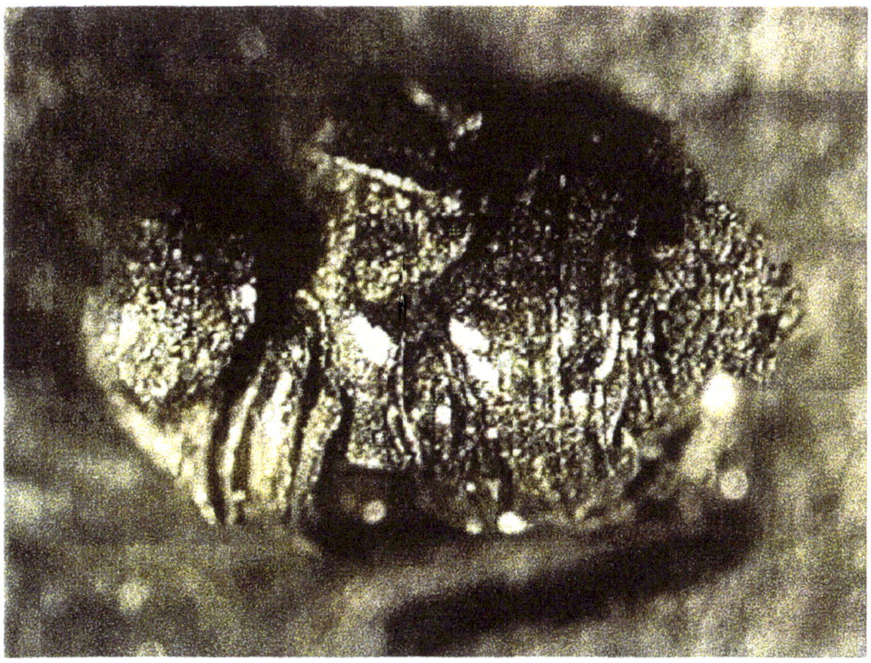

Pl. 2. The charred larva of a bruchid beetle, *Bruchus rufipes*, recovered from pulses charred by the eruption, Akrotiri, Thera, Greece (L. 2mm.).

Pl. 3. Virtually complete and uncharred specimen of the lesser grain borer beetle, *Rhyzopertha dominica* from Akrotiri, Thera, Greece. Dorsal view (L. 2.5mm.).

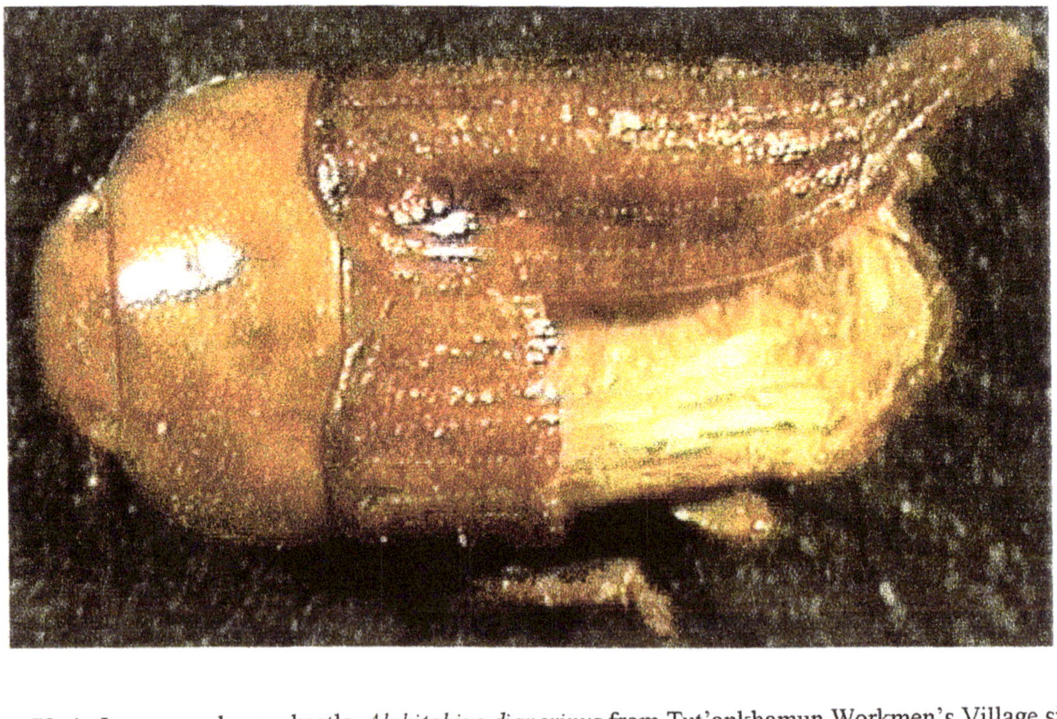

Pl. 4. Lesser mealworm beetle, *Alphitobius diaperinus* from Tut'ankhamun Workmen's Village site, Amarna, Egypt.

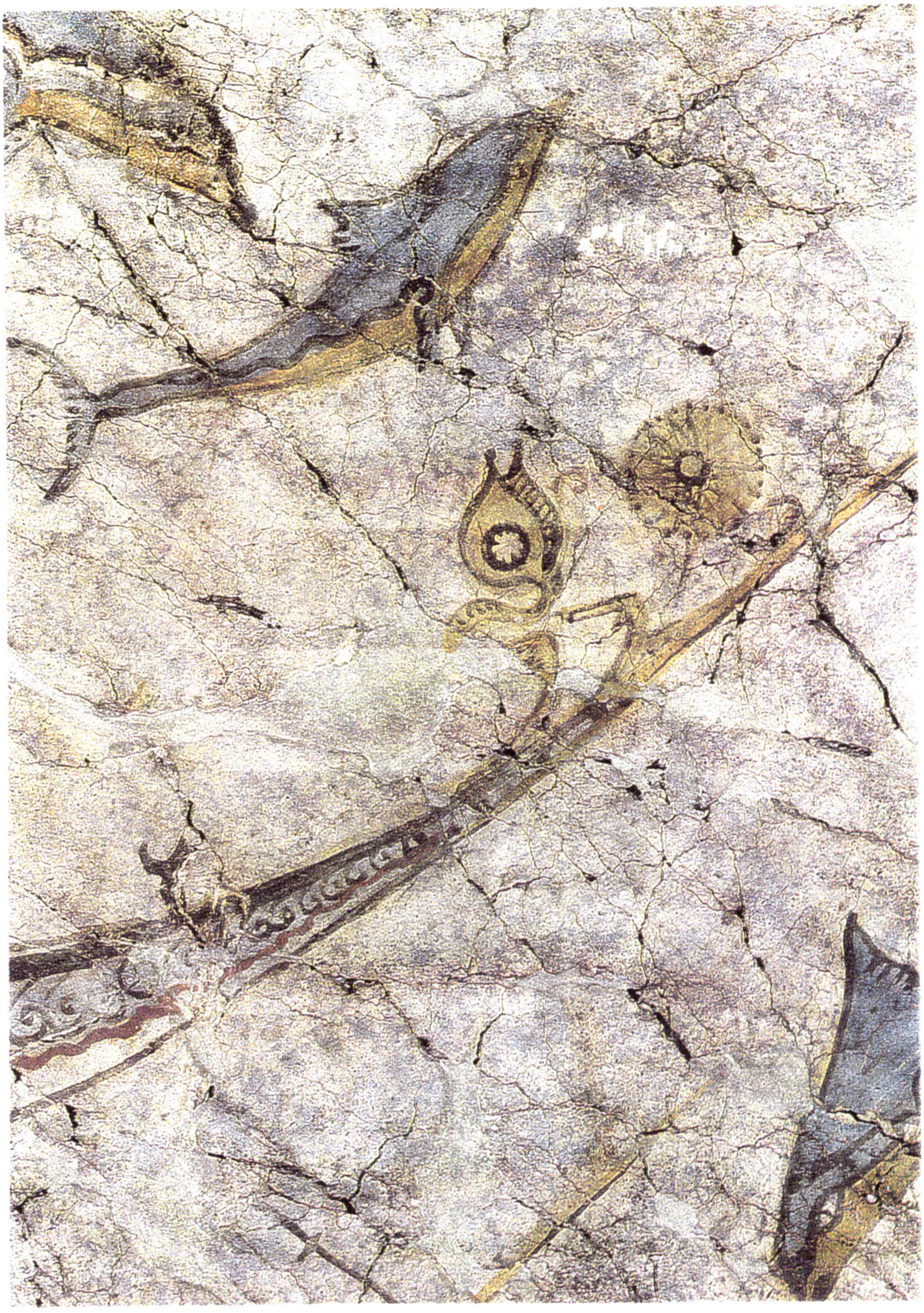

Pl. 5. Detail of the flotilla fresco, West House, Akrotiri, Thera, Greece, showing lepidopteran interpreted as silk moths, and flowers, perhaps of cotton on the prow of Ship 2 (after Doumas 1992).

Pl. 6. Xeste 3, 'Lustral basin' fresco on North Wall. Detail from the 'Adorants', showing woman in diaphonous garments (after Doumas 1992).

www.ingramcontent.com/pod-product-compliance
Lightning Source LLC
LaVergne TN
LVHW071215210926
848482LV00015B/814